Information Circular 9510

A Performance Evaluation of Two Overhead Power Line Proximity Warning Devices

By Gerald T. Homce, P.E., James C. Cawley, P.E., and Michael R. Yenchek, P.E.

DEPARTMENT OF HEALTH AND HUMAN SERVICES
Centers for Disease Control and Prevention
National Institute for Occupational Safety and Health
Pittsburgh Research Laboratory
Pittsburgh, PA

November 2008

This document is in the public domain and may be freely copied or reprinted.

Disclaimer

Mention of any company or product does not constitute endorsement by the National Institute for Occupational Safety and Health (NIOSH). In addition, citations to Web sites external to NIOSH do not constitute NIOSH endorsement of the sponsoring organizations or their programs or products. Furthermore, NIOSH is not responsible for the content of these Web sites. All Web addresses referenced in this document were accessible as of the publication date.

Ordering Information

To receive documents or other information about occupational safety and health topics, contact NIOSH at

>Telephone: **1–800–CDC–INFO** (1–800–232–4636)
>TTY: 1–888–232–6348
>e-mail: cdcinfo@cdc.gov
>
>or visit the NIOSH Web site at **www.cdc.gov/niosh**.

For a monthly update on news at NIOSH, subscribe to NIOSH *eNews* by visiting **www.cdc.gov/niosh/eNews**.

DHHS (NIOSH) Publication No. 2009–110

November 2008

SAFER • HEALTHIER • PEOPLE™

CONTENTS

Page

Executive summary 1
Introduction 2
Background 3
    Detection of energized overhead power lines 3
    Proximity warning device descriptions 3
        SIGALARM model 210 3
        Allied Safety Engineering model 2100 5
    Proximity warning device performance objective 6
Test program 7
    Installation of power line proximity warning devices on the test crane 7
    Test facility 8
        Overhead power line test site 8
        Electrical power supply for tests 9
        Data collection process 10
    Test procedures 11
    Crane boom position data analysis 13
    Test plan 13
Results 14
    Test set 1 14
    Test set 2 16
    Test set 3 17
    Test set 4 19
    Test set 5 21
    Test set 6 23
    Test set 7 25
    Test set 8 27
    Test set 9 28
    Test set 10 28
Summary of proximity warning device performance 29
    Power line configurations 29
    Crane boom configurations 32
    Other factors 34
        Use of a PWD as an early warning device 34
        Effect of connecting the crane chassis to a grounded power system neutral 34
        Effect of multiple independent power line circuits 34
        Effect of power line phase sequence 34
    Sensitivity adjustments 35
Recommendations for future research 36
References 36
Appendix A.—List of cooperators 37

CONTENTS—Continued

ILLUSTRATIONS

Page

1. SIGALARM model 210 control unit 4
2. ASE model 2100 remote and operator panels..5
3. Minimum safe working distances around overhead electrical power lines 6
4. PWD probes on crane boom 8
5. Diagram of the test-site power system..9
6. Example of a computer model used for analyzing crane boom position relative to overhead power lines 10
7. Crane locations for tests involving operation from a stationary position 11
8. Grove RT522 crane on outriggers, positioned adjacent to overhead power lines with boom at 50% extension and boom tip approximately level with lower set of lines 12
9. Test set 1 layout and overhead power line configuration 15
10. Test set 2 layout and overhead power line configuration 16
11. Test set 3 layout and overhead power line configuration 18
12. Test set 4 layout and overhead power line configuration 19
13. Test set 5 layout and overhead power line configuration 21
14. Test set 6 layout and overhead power line configuration 23
15. Test set 7 layout and overhead power line configuration 25
16. Test set 8 layout and overhead power line configuration 27
17. Test set 9 layout and overhead power line configuration 28

TABLES

1. Power line configuration detail for test sets 1 through 7 13
2. Alarm distance results for test set 1 15
3. Alarm distance results for test set 2 17
4. Alarm distance results for test set 3 18
5. Alarm distance results for test set 4 20
6. Alarm distance results for test set 5 22
7. Alarm distance results for test set 6 24
8. Alarm distance results for test set 7 26
9. Summary of PWD alarm performance based on power line configuration............................30
10. Summary of PWD alarm performance based on power line configuration when analysis is limited only to tests where the boom extension was at 50% and the boom elevation was such that the tip was approximately at power line level............................31
11. Summary of PWD alarm performance based on crane boom elevation angle or extension 33
12. Summary of PWD alarm performance changes due to reversal of power line phase sequence 35

ACRONYMS AND ABBREVIATIONS USED IN THIS REPORT

3-D	three-dimensional
ACSR	aluminum conductor steel-reinforced
ANSI	American National Standards Institute
ASME	American Society of Mechanical Engineers
AWG	American wire gauge
CFR	Code of Federal Regulations
LCD	liquid crystal display
NIOSH	National Institute for Occupational Safety and Health
OSHA	Occupational Safety and Health Administration
PRL	Pittsburgh Research Laboratory (NIOSH)
PWD	proximity warning device

UNIT OF MEASURE ABBREVIATIONS USED IN THIS REPORT

cm	centimeter
ft	foot
ha	hectare
Hz	hertz
in	inch
kV	kilovolt
kVA	kilovolt-ampere
m	meter
min	minute
mt	metric ton
st	short ton
V	volt
W	watt
Ω	ohm

A PERFORMANCE EVALUATION OF TWO OVERHEAD POWER LINE PROXIMITY WARNING DEVICES

By Gerald T. Homce, P.E.,[1] James C. Cawley, P.E.,[2] and Michael R. Yenchek, P.E.[3]

EXECUTIVE SUMMARY

Accidental contact of overhead electrical power lines by mobile equipment is a leading cause of occupational fatalities in the United States, accounting for 20% of on-the-job electrocutions. Overhead electrical power line proximity warning devices (PWDs) are intended to warn personnel if mobile equipment moves within some preselected minimum distance of an energized overhead electrical power line. Two commercially available PWDs were tested at the National Institute for Occupational Safety and Health's (NIOSH) Pittsburgh Research Laboratory (PRL). The objective of the tests was to document performance capabilities and limitations for these PWDs by identifying factors that can influence their operation.

The two PWDs evaluated in this research are the SIGALARM Model 210 marketed by Allied Safety Systems, LLC, and the ASE Model 2100 from Allied Safety Engineering. Both of these devices operate by measuring the electric field present around energized power lines. The PWDs were installed on a government-owned 22-st (20-mt) rough terrain crane. A purpose-built test site used for this research at PRL allowed operation of the crane near a variety of power line configurations operating at up to 25 kV. Most of the tests involved positioning the crane adjacent to one or more overhead power lines, adjusting sensitivities of the PWDs to alarm when the crane boom was approximately 20 ft (6.1 m) from the power lines, swinging the crane boom toward the lines under a wide variety of test conditions, and finally, for each unique set of test conditions, documenting the deviation from 20 ft (6.1 m) for actual alarm activation.

Test results show that several factors can adversely affect PWD performance. PWD alarm accuracy generally deteriorated when operating with a boom position significantly different than that used for the device's last sensitivity adjustment. Another factor that can affect PWD performance is configuration of the overhead power line(s) involved. Accuracy of alarm activation distances was best for simple single-circuit installations, but degraded for multiple circuits on the same poles. This degradation was slightly greater for installations with different voltage levels and/or a combination of vertical and horizontal conductor arrangements. Performance also degraded for crane operation between two intersecting power line installations, especially for intersecting lines at different voltages. An additional aspect of power line configuration shown to influence PWD accuracy was phase sequence on the power line circuit(s). Specific phase conductor arrangements and combinations, particularly in multiple circuit installations, resulted in either improved or degraded accuracy.

---

[1]Lead Research Engineer, Electrical.
[2]Senior Research Engineer, Electrical (retired).
[3]Lead Research Engineer, Electrical.
Pittsburgh Research Laboratory, National Institute for Occupational Safety and Health, Pittsburgh, PA.

Tests were also conducted to evaluate the PWDs as "early warning devices" for situations such as moving a mobile crane into an unfamiliar work area. Results showed that the SIGALARM Model 210 could detect energized 13-kV power lines at a distance of 75–88 ft (22.9–26.8 m). This alarm distance would allow an operator to take preventive measures before the crane is in a position from which it could contact nearby power lines.

INTRODUCTION

Many of the electrical fatalities in construction, mining, and other industries are due to personnel accidentally contacting overhead electrical power lines with high-reaching equipment such as mobile cranes. During a recent 10-year period, approximately 20% of occupational electrocutions involved contact between mobile equipment and overhead power lines [Cawley and Homce 2003]. In a typical power line contact accident, the frame of the equipment (and possibly a suspended load in the case of mobile cranes) is energized to a high voltage relative to the surrounding ground surface. Anyone touching the frame and ground simultaneously is exposed to this high voltage and can become a path for lethal levels of electrical current.

Overhead electrical power line PWDs are mobile equipment-mounted safety devices intended to alert personnel if the equipment is operating too close to an energized overhead electrical power line. Such devices have been commercially available for more than 30 years, but have not found widespread acceptance in many industries due, in part, to a lack of regulatory requirements for their use. The Occupational Safety and Health Administration (OSHA) is currently involved in updating the standards for cranes and derricks (29 CFR[4] 1926.550). Part of the proposed revision addresses overhead power line safety for mobile cranes and includes explicit reference to PWDs as one of several acceptable measures for protecting workers from accidental power line contacts. With this proposal to accept PWDs as one means to maintain a safe distance between cranes and power lines (as specified in 29 CFR), NIOSH researchers concluded that an objective performance evaluation of PWDs would be valuable and timely.

A performance evaluation of two commercially available overhead power line PWDs was conducted at NIOSH–PRL. The objective of the tests was to document performance capabilities and limitations for these PWDs by identifying factors that can influence their operation. The overall approach for this testing called for the two PWD companies to install their devices on a government-owned 22-st (20-mt) rough terrain crane and specify procedures for their use. The crane was to be operated using a wide range of boom positions near several different configurations of energized overhead power lines, with the performance of the PWDs documented. This full-scale testing took place at a purpose-built overhead power line test site at PRL. PRL engineers coordinated and directed this research, but input for developing the test protocol was solicited from a number of cooperators, including the two PWD manufacturers participating in the study, an equipment manufacturing trade association representative, labor union representatives, OSHA, a large private construction and crane rental firm with experience using PWDs, and an electrical engineering consulting firm working as a NIOSH contractor. A list of these participants can be found in Appendix A.

---

[4] *Code of Federal Regulations.* See CFR in references.

BACKGROUND

Detection of Energized Overhead Power Lines

Energized power lines radiate both magnetic and electric fields. Therefore, the presence of energized power lines may be detected by measuring either the magnetic field or the electric field surrounding them. The magnetic field strength at any point around power lines is, in part, a function of the electrical current being carried by the lines. Since power line current can vary significantly, magnetic field measurement cannot reliably be used to determine distance to power lines. In contrast, the electric field surrounding power lines is primarily a function of the voltage at which they operate and their configuration.[5] Therefore, the key parameters that determine electric field strength at any given position near a set of conductors are fixed for a given power line installation, making electric field measurement more suitable for estimating the distance to energized power lines [Hipp et al. 1980].

Proximity Warning Device Descriptions

Two commercially available PWDs were evaluated in this study. Each is designed to warn of proximity to power lines by detecting the electric field present around the phase conductors. In general, each device consists of a probe, a control unit, and an audible alarm. The probes were deployed along the length of the boom, and the control unit was installed in the crane cab, accessible to the operator. The control unit for each device includes a sensitivity adjustment and a user interface for alarm and PWD status information. Each device measures the 60-Hz signal derived from the probe exposed to the power line electric field. The resulting signal level is compared to the alarm threshold preselected by the operator; if it is exceeded, audible and visible alarms are activated. If the field measurement level falls below the threshold level, the alarm ceases. The general procedure for adjusting the alarm threshold or "sensitivity" of each device is to position the crane boom at the desired minimum working distance from the power lines and adjust the device to establish an alarm threshold at that boom position. The two PWDs are described in more detail below.

SIGALARM Model 210

The SIGALARM Model 210, marketed by Allied Safety Systems, LLC, is a power line PWD designed for use on telescoping boom (hydraulic) cranes [Allied Safety Systems, LLC 2008]. Figure 1 shows the SIGALARM Model 210 control unit. This model uses an automatically retracting reel for the probe to accommodate the extension and retraction of the boom and can also be equipped with additional alert devices. The probe is a two-conductor insulated cable, with one conductor used as the antenna and the other as a path through which to inject a device test signal at the end of the antenna (using a test button on the control module). The control module mounts in the crane cab for operator access. It has rotary switches for coarse and fine

---

[5]For a specific point in space near overhead power lines, the electric field will vary depending on factors such as vertical versus horizontal phase conductor arrangement, conductor spacing, the presence of multiple power line circuits, and the presence of a neutral conductor.

sensitivity adjustment (referred to as "calibration" in the Model 210 manual), power and test switches, and status indicator lights.[6]

In operation, the Model 210 automatically adjusts to its maximum sensitivity when powered on. The sensitivity is set to the desired level by first adjusting the coarse and then fine sensitivity adjustments, and when in alarm condition, the unit emits a loud, repetitive chirping signal. When calibrated by the manufacturer at 20 ft (6.1 m) from the power line during testing, the alarm rate was approximately two chirps per second. When well beyond 20 ft (6.1 m) from the power lines, the Model 210 produced no chirping. As the electric field strength increased during boom movement, the chirp began and its rate increased. A rate of about two chirps per second indicated that the field strength measured by the device was equivalent to that during calibration.

Figure 1.—SIGALARM Model 210 control unit.

---

[6]Installation of the Model 210 for this research comprised an unshielded, jacketed twisted pair cable for the probe. When not serving as a means to inject a test signal, the second conductor in the probe is simply an extension of the antenna conductor, since it is connected to the antenna conductor at the tip of the crane boom. The Model 210 can use any of several different cable types as a probe, as described in the installation manual [Allied Safety Systems, LLC 2008].

Allied Safety Engineering Model 2100

The ASE Model 2100 was a microprocessor-based power line PWD developed by Allied Safety Engineering. The device tested was a preproduction unit very similar to the Model 2200 on Allied Safety Engineering's Web site at the time of this writing [Allied Safety Engineering 2008]. The ASE Model 2100 consisted of a cab-mounted remote panel (main control module) to which two probes (antennas) and an operator panel were connected. The unit could drive multiple external alert devices. The probes consisted of two insulated conductors positioned one on each side of the boom and deployed from two automatically retracting reels. The operator panel provided an LCD display of field strength, sensitivity adjustments, alarm status, and other system information, as well as several pushbuttons to allow operator input and control. Figure 2 shows the Model 2100 remote and operator panels.

Sensitivity was adjusted by selecting a field strength "set point" when the boom was positioned at the desired distance from the power lines. The set point and the relative field strength at the device probes were continuously displayed on the operator panel as dimensionless whole numbers scaled for easy interpretation by the operator. Two audible alarms—a warning signal and an alarm signal—were annunciated as a series of beeps followed by a voice message. At 80% of the alarm set point (this percentage is a user-selectable value), a "warning" voice message was given indicating that continued boom movement in the same direction may result in an alarm condition. The "alarm" voice message warned of impending power line contact.

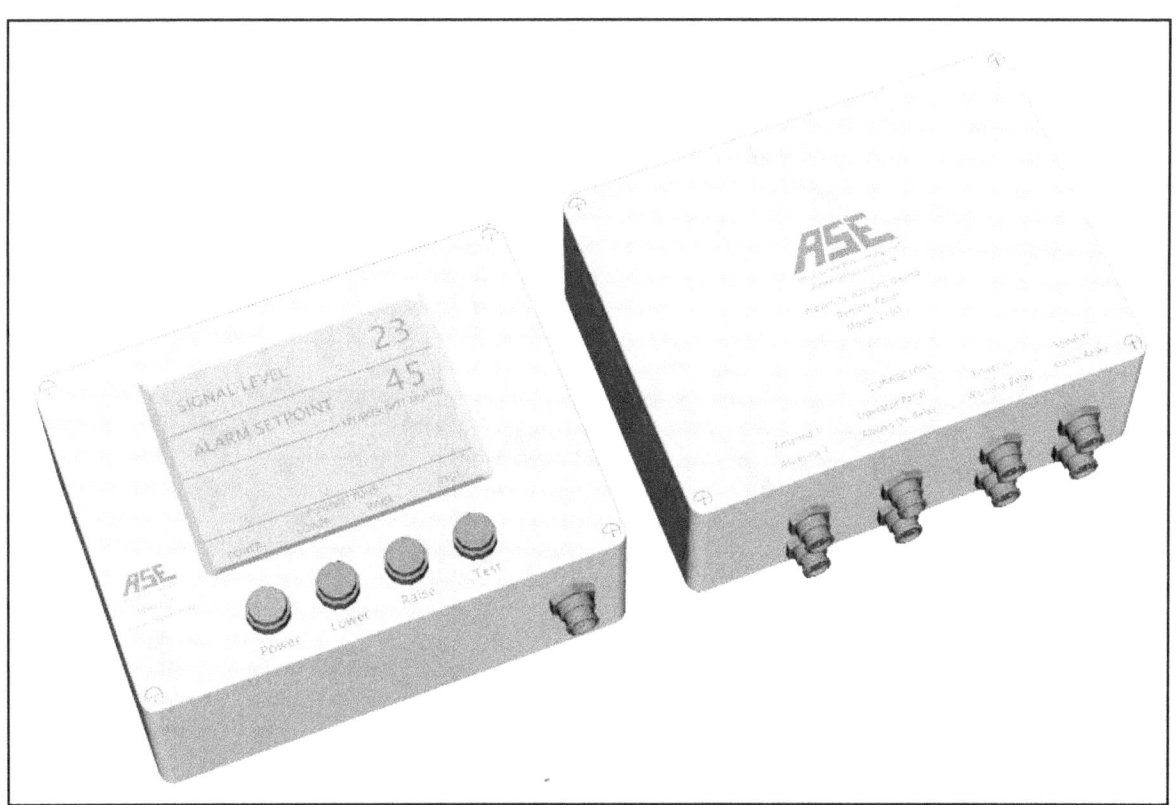

Figure 2.—ASE Model 2100 remote and operator panels.

Proximity Warning Device Performance Objective

In the case of a mobile crane such as the one used in this research, a PWD should ideally warn the equipment operator and any nearby personnel when any part of the crane or a suspended load is closer to an energized overhead electrical power line than some preselected distance. In most cases, the minimum distance will be dictated by occupational safety regulations, such as the current OSHA standards that require a clearance of at least 10 ft (3.0 m) for power lines at voltages up to 50 kV (29 CFR 1926.550). Conversely, a PWD should not alarm during operation outside the desired minimum clearance distance. Such "false alarms" may cause personnel to question the effectiveness of the device, view it as a nuisance, or perhaps ignore it completely. The criteria just described, strictly speaking, would allow operation of a crane boom directly above or below a power line as long as the specified minimum clearance is maintained. Such operation, however, increases the chance of an accidental line contact and should be avoided [ANSI 1995]. A safer and more practical alternative is to define the limit of safe operation as a vertical plane at the desired clearance distance horizontally from the closest power line conductor and prohibit any part of the crane or suspended load from crossing this boundary during normal operation. Figure 3 shows a three-phase power line installation in cross-section and two variations of a 10-ft (3.0-m) recommended safe operation boundary surrounding the conductors.

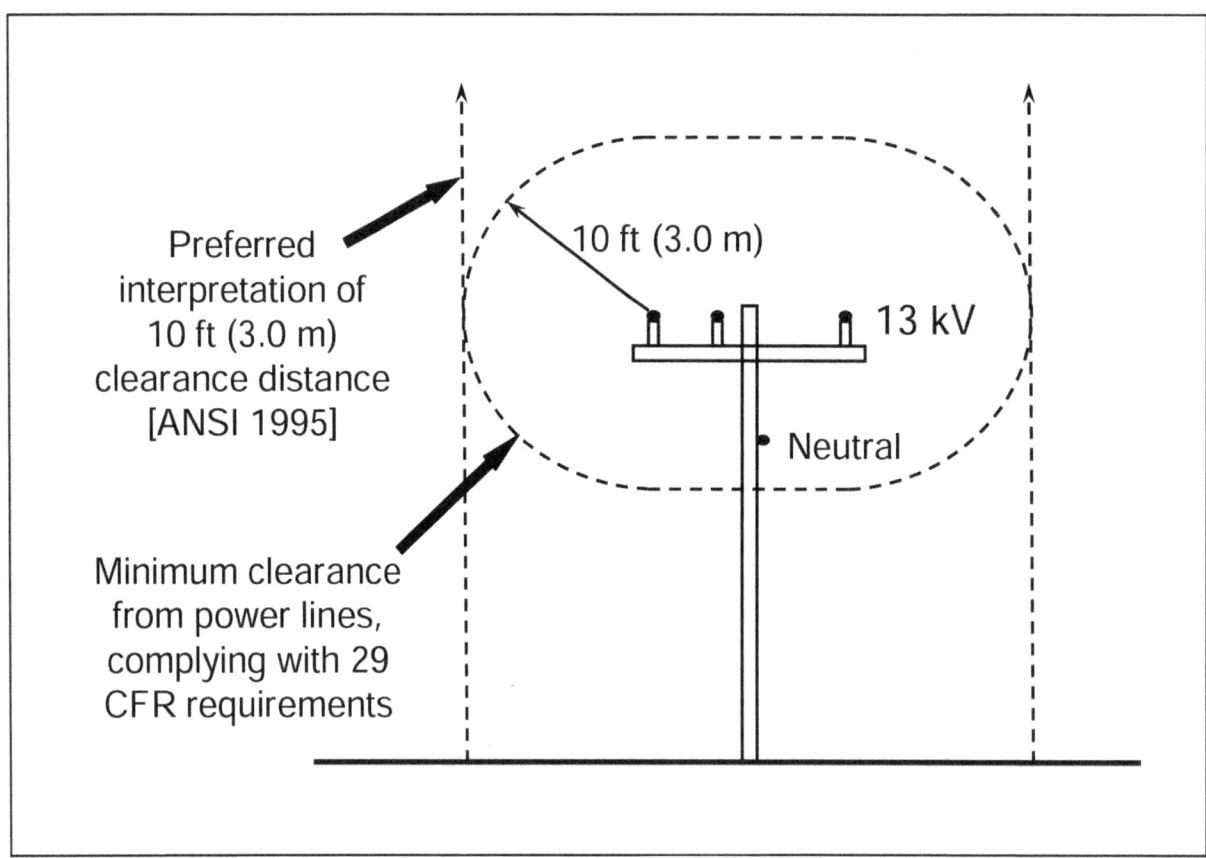

Figure 3.—Minimum safe working distances around overhead electrical power lines.

For most tests in this evaluation, a minimum clearance distance of 20 ft (6.1 m) from the closest phase conductor of the power lines was used as the limit of safe operation and, accordingly, the PWDs had their sensitivities adjusted so that their alarm thresholds were reached with the boom tip approximately 20 ft (6.1 m) from the closest phase conductor.[7] This distance was chosen to allow the crane boom to continue moving toward the power lines during testing if an alarm did not activate at or before the 20-ft (6.1-m) limit, without bringing any part of the crane closer than 10 ft (3.0 m) from the lines. This boundary of safe operation could have been defined as a vertical plane 20 ft (6.1 m) from the power lines. The more literal interpretation of the boundary was used, however, allowing a minimum distance of 20 ft (6.1 m) in all directions from any energized phase conductor. The main implication for the performance evaluations presented in this report is that movement of the boom below the power lines (at 0° elevation angle), without activating the alarms, is considered acceptable PWD operation. The exceptions to the 20-ft (6.1-m) sensitivity setting were the tests involving the crane traveling toward the lines with the PWDs set at maximum sensitivity.

TEST PROGRAM

Installation of Power Line Proximity Warning Devices on the Test Crane

Both PWDs were mounted by their manufacturers on a government-owned Grove RT522 rough-terrain crane (22-st (20-mt) capacity). Typical mounting techniques and routing of wiring were altered to avoid permanent modifications to the crane, but proper probe placement was maintained.

Operator control panels for both units were located in the crane cab, and speakers for audible alarms were located on the base section of the boom. The SIGALARM Model 210 probe reel was mounted on the lower half of the boom base section, and the probe routed to the boom tip on standoffs several inches above the upper left edge of the boom. Probe extensions approximately 24 in (61 cm) long were mounted to hang freely at the boom tip to provide extra coverage in that area. The two probes for the ASE Model 2100 were extended from two reels mounted on both sides of the base section of the boom and ran along each lower outside edge of the boom to the boom tip. Figure 4 shows the PWD probes mounted on the crane boom.

---

[7]For this sensitivity adjustment distance, the boom was positioned by placing the crane load block directly over a mark painted on the ground 20 ft (6.1 m) horizontally from the closest conductor. Because of the inaccuracy inherent in this method and the size and shape of the boom structure, the actual minimum distance from the boom to the closest conductor (as subsequently measured by computer model) was usually not precisely 20 ft (6.1 m). The 20-ft (6.1-m) "nominal" distance, however, was used as a reference for all evaluations since this is the target value that would be assumed as correct by a crane operator using a similar procedure to adjust PWD sensitivity in the field.

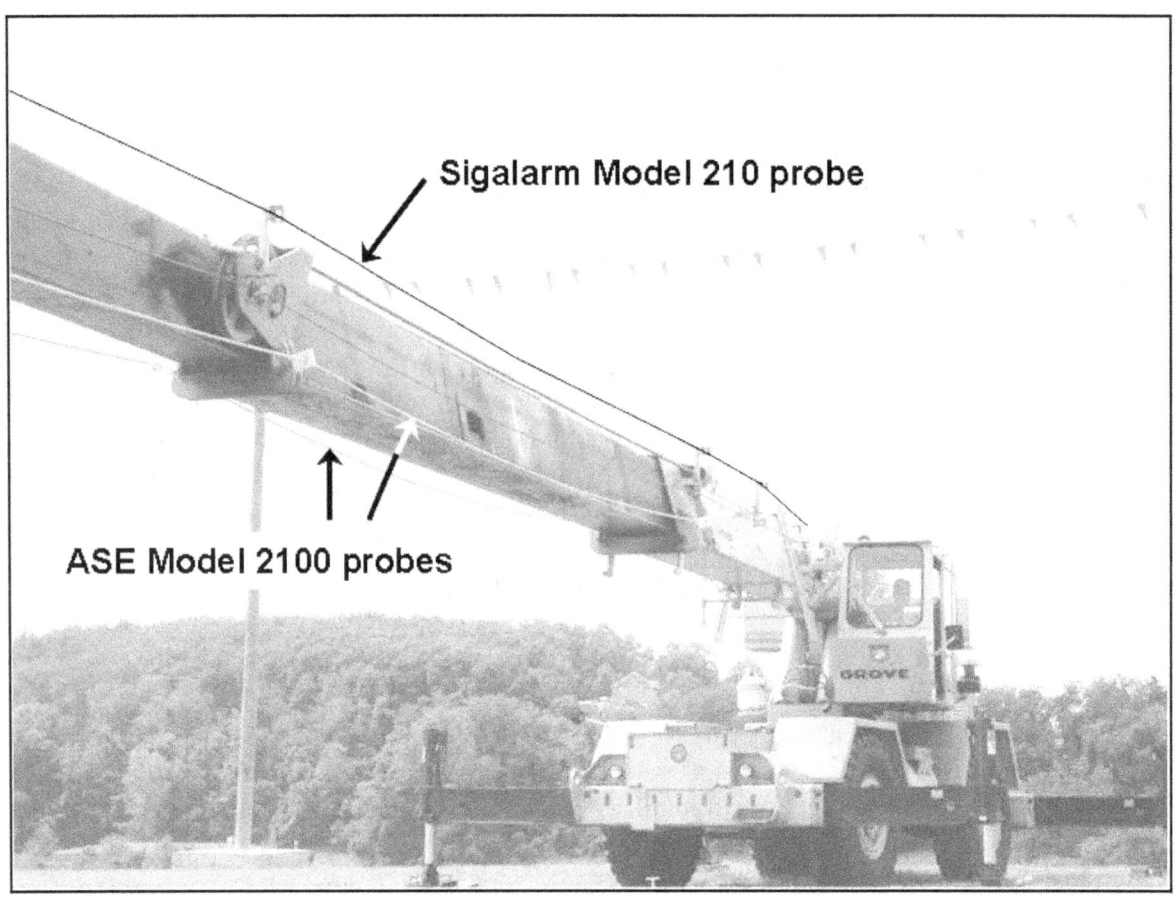

Figure 4.—PWD probes on crane boom.

Test Facility

Overhead Power Line Test Site

A test site specifically designed for operating mobile equipment near overhead electrical power lines was built at NIOSH–PRL. The site is a remote, approximately level, 5-acre (2-ha), grass-covered area with controlled access. Within this area, four wooden utility poles were erected along a straight line, 100 ft (30.5 m) apart, and equipped with cross arms and necessary hardware to install one or two four-wire three-phase horizontal configuration power line circuits and a three-wire vertical configuration circuit. The lines can be installed in various combinations at heights up to 45 ft (13.7 m) and can be powered from either end of the 300-ft (91.4-m) installation. Four "movable" wooden utility poles were also erected at the site. These 36-ft (11.0-m) poles set in 5.5-st (5-mt) reinforced concrete blocks are designed to be moved by a forklift and can be equipped with one four-wire three-phase horizontal configuration circuit. The movable poles can be positioned to create a power line circuit parallel to, or intersecting, the lines on the four permanent poles. For both sets of poles, all insulators and cross arms can accommodate circuits up to 25 kV using #2 AWG ACSR conductors. These overhead power line installations meet standards used by Allegheny Energy, Inc., a large electrical utility company headquartered near Pittsburgh, serving Pennsylvania, Maryland, West Virginia, and Virginia. A nonstandard safety feature included on the utility poles is a nonconductive, high-visibility flag line suspended level with, and 10 ft (3.0 m) away from, the power lines (on the side of crane operation). This gives the crane operator and spotters a clear indication when the crane boom or hoist ropes are 10 ft (3.0 m) horizontally from the power line conductors. The site also has a dedicated low-resistance (3 Ω), driven-rod electrical ground bed. When the overhead power lines were configured for a specific test, any conductors not in use were removed from the utility poles.

Electrical Power Supply for Tests

A portable electrical power supply was assembled to supply overhead power line circuits at the power line test site described above. It can provide two independently controlled four-wire three-phase sources, isolated from utility sources, and is capable of 27 kV line to line. The system uses a 480-V, three-phase, 5,000-W generator feeding two parallel branches. Each branch has a three-phase variable transformer capable of providing a manually adjustable 0- to 560-V output, feeding a 480-V to 27-kV, 2-kVA step-up transformer. Each branch can serve as an independent 0- to 27-kV four-wire three-phase supply for an overhead power line circuit. Four-conductor insulated portable power cables terminated with four-prong twist-lock plugs connect all 480-V components. High-voltage outputs use high-voltage insulated conductors connected to power line terminations on cross arms 10 ft (3.0 m) above ground level. A circuit breaker at the generator output and fuses at transformer primaries provide overcurrent protection for the system. All components can be moved from storage and setup for operation by a small team in 30–40 min. Figure 5 is a diagram of the power system.

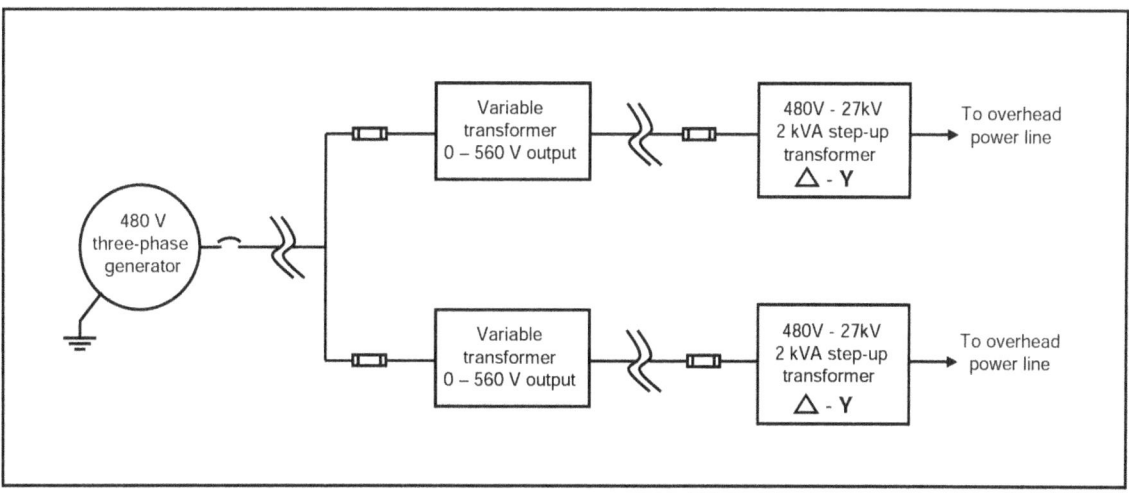

Figure 5.—Diagram of the test-site power system.

For this test program, the output for each power line circuit was monitored at the step-up transformer secondary terminals using a digital multimeter and high-voltage probe to measure one phase to neutral voltage. For personnel safety, access to the areas surrounding the step-up transformers and high-voltage connections at ground level were restricted using construction fencing. All tests using a four-wire circuit were configured with the high-voltage output grounding conductor tied to the test site ground bed and overhead line neutral conductor, as well as a butt-wrapped grounding lead on each pole. Prior to testing, all line-to-line voltages were verified directly on the overhead lines using a hot-stick meter employed from an aerial manlift.

Data Collection Process

Data collection for this research required documenting the crane boom position relative to the energized overhead power lines for each test in order to allow evaluation of PWD performance. This was done by determining the relative three-dimensional (3-D) positions of the power lines and boom and recording this information as a computer model from which the relevant boom to power line distances ("true" distances) can be extracted. The 3-D information was obtained using a "total station" surveying instrument, which allows precise measurement of horizontal and vertical angles as well as slope distance from the instrument to target points. Use of the instrument required precise placement of reference targets (made of a special reflective material) on the crane boom and the overhead power line installations.

Using the survey coordinates of two reference points on either (or both) sets of overhead power lines and accurate measurements of the poles and conductors, a 3-D model of the test site for each power line configuration was created. Then for all tests, the boom positions, for both sensitivity adjustments and cases of alarm activation, were mapped in the same 3-D space as the power lines. Each boom model was based on the survey coordinates of two reference points on the boom and detailed boom dimensions. Figure 6 is an example of a computer model for an individual test. With the graphical computer models of the power lines and crane boom/hoist ropes mapped in the same 3-D space, the distance between the boom/hoist ropes and power lines is readily determined by simply defining the shortest line between potential contact points for the two and noting its length.

To verify the boom to power line distances obtained from the models, manual measurements of the horizontal boom to power line distances ("observed" distances) were made for all alarm activations. These were made by noting the position of the crane load hook with respect to a distance reference grid painted on the ground at the test site. The zero lines for the distance reference grid were directly below the nearest phase conductors for horizontal configuration lines (at 13-kV spacing).

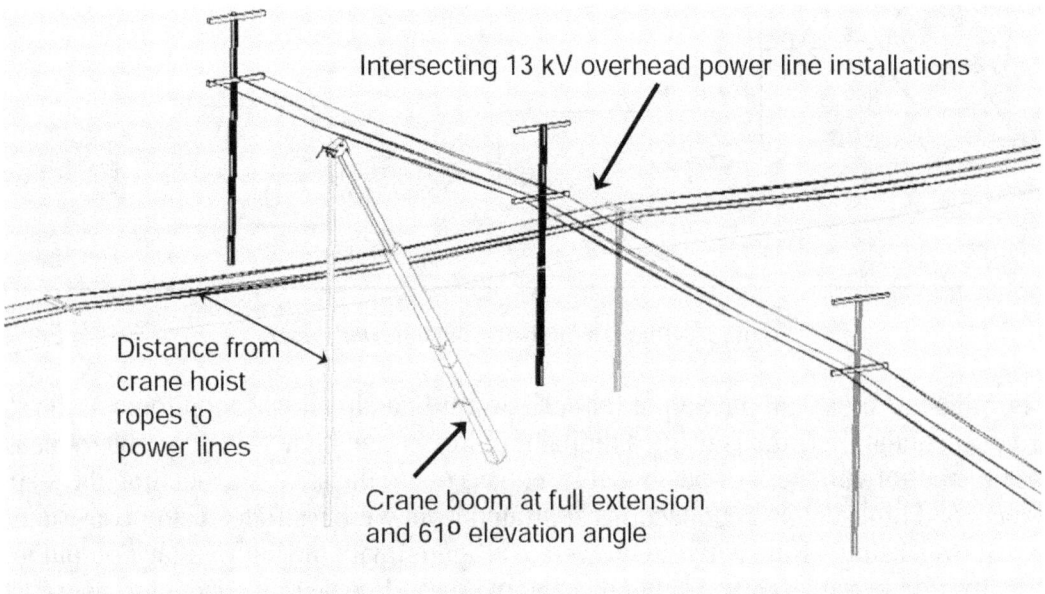

Figure 6.—Example of a computer model used for analyzing crane boom position relative to overhead power lines.

Test Procedures

The PWDs, as mounted on the crane by the manufacturers, were tested by operating the crane near energized overhead power lines. Different test sets varied the crane boom position relative to the power lines, as well as power line configuration (number of three-phase circuits, pole configuration, conductor arrangement, voltage level, and phase sequencing). Within each test set, individual tests involved swinging the boom toward the lines while at predetermined combinations of crane boom extension and elevation angle, and then documenting the response of each PWD. Most tests specified that the crane be operated from a stationary position (chassis supported on outriggers). However, several tests involved the crane approaching the power lines to simulate entering an unfamiliar work area.

A typical test, for operation of the crane from a stationary position, began by positioning the crane adjacent to the power lines and deploying its outriggers, supported on wooden blocks. For tests using a single set of poles, the crane chassis centerline was parallel to the power lines and approximately 30 ft (9.1 m) horizontally from the closest phase conductor, as shown in Figure 7(a). For two sets of poles with power lines intersecting at 90°, the crane chassis centerline bisected the angle between the power lines, with the boom's center of rotation approximately 30 ft (9.1 m) horizontally from the closest phase conductor of each set, as shown in Figure 7(b).

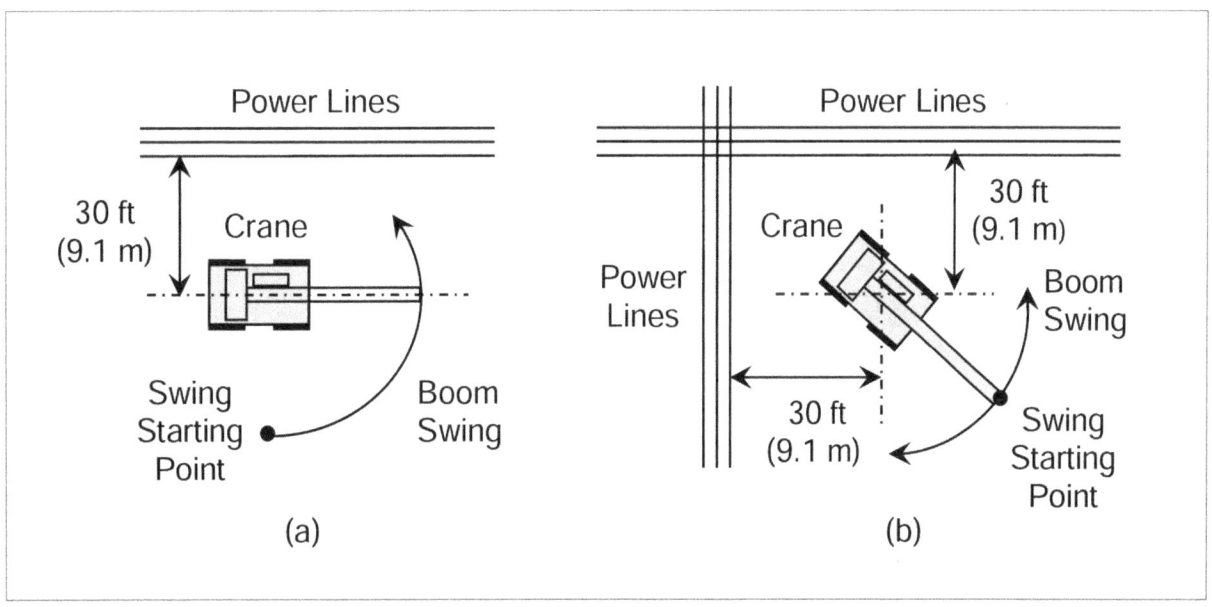

Figure 7.—Crane locations for tests involving operation from a stationary position.

PWD sensitivities were then adjusted per on-site direction from the manufacturers. The criteria used for these adjustments called for the PWDs to alarm if the crane boom or hoist ropes moved to within 20 ft (6.1 m) of the closest phase conductor. After sensitivity adjustment, the boom's 3-D position relative to the power lines was documented using the total station surveying instrument. Notwithstanding the 20-ft (6.1-m) alarm requirement, the manufacturers were free to specify the specific procedures for adjusting PWD sensitivities, as well as how frequently the devices would be readjusted during testing.

The crane boom was then rotated horizontally to the swing starting point as shown in Figure 7, elevated and extended per the test plan, and swung toward the power lines. When either PWD alarm would activate, the boom was stopped and its 3-D position relative to the power lines was documented. The observed distance from the boom or hoist ropes to the power lines was also estimated using a distance reference grid painted on the ground. If an alarm activated at the maximum distance from the power lines (i.e., before the boom was moved), this position was documented. If a PWD did not alarm at or before the 20-ft (6.1-m) limit, the boom swing was continued until alarm activation occurred or until the boom or hoist ropes were 10 ft (3.0 m) from the power lines. The boom and hoist ropes were never allowed to come closer than 10 ft (3.0 m) to an energized power line during testing. Note that in cases where the boom elevation was 0°, the boom was swung until directly under the closest phase conductor, since at this point it was still greater than 20 ft (6.1 m) (vertically) from the power lines. Figure 8 shows the test site and crane.

Figure 8.—Grove RT522 crane on outriggers, positioned adjacent to overhead power lines with boom at 50% extension and boom tip approximately level with lower set of lines.

For tests that simulated the crane, in travel configuration, approaching the power lines, the PWDs were adjusted to maximum sensitivity. Upon activation of an alarm, the crane was stopped and set up for stationary operation on outriggers at that location. Tests were then conducted by swinging the fully extended boom toward the power lines while at different elevations.

Crane Boom Position Data Analysis

Position survey data and computer modeling were used to precisely record the relative positions of the overhead power lines and crane boom for sensitivity adjustment positions and each alarm activation point during testing. The criteria used to determine the true distance between the boom or hoist ropes and the power lines for each test (from the models) was based on the position of the boom and the most likely point of contact with a power line conductor. For tests where the boom was at a 0° elevation angle or where the boom tip was at power line level, the true distance is the shortest distance from the tip of the boom to the closest phase conductor. For tests where the boom elevation angle was at the maximum used (61°), the true distance is the shortest distance from the hoist ropes to the closest phase conductor. Consideration of power line contact by hoisted loads was beyond the scope of this research.

Test Plan

Most of the tests conducted in this research quantified the influence on PWD performance of variations in power line configuration and changes in crane boom parameters after initial PWD sensitivity adjustment. Table 1 is a listing of test sets 1 through 7 organized according to power line configuration (crane operating from a stationary position).

Table 1.—Power line configuration detail for test sets 1 through 7

Test set No.	No. of three-phase circuits	Circuit No. 1			Circuit No. 2			Relationship of two power line circuits
		Voltage (kV)	Configuration	Neutral conductor?	Voltage (kV)	Configuration	Neutral conductor?	
1	1	13	H	Yes				
2	1	25	V	No[1]				
3	2	13	H	Yes	13	H	Yes	On same poles
4	2	13	H	Yes	25	H	Yes	On same poles
5	2	13	H	Yes	25	V	No[1]	On same poles
6	2	13	H	Yes	13	H	Yes	Intersect at 90°
7	2	13	H	Yes	25	H	Yes	Intersect at 90°

H Horizontal.　V Vertical.
[1]This 25-kV power line simulates a subtransmission circuit, which typically would not include a neutral conductor.

Each test set was run once with the phase conductors arranged ABC with A-phase nearest the crane for horizontal lines and at the top for vertical lines, and then run again with the circuit (or one of the two circuits involved) arranged CBA in order to evaluate the effect of reversing the electrical "sequence" of the phase conductors. Within each of the 14 resulting groups of tests, an individual test was run for each of 8 possible crane boom configurations. These configurations represented all possible combinations of three boom extension lengths (minimum, 50%, and maximum)[8] and three elevation angles (0°, boom tip at power line height, and 61°).[9,10] The crane load block (hoisting hook) was positioned approximately 5 ft (1.5 m) from the ground for sensitivity adjustments and tests. The exception to the 5-ft (1.5-m) load block height was for the combination of maximum boom extension and 61° elevation angle, for which the hoist block was positioned 23 ft (7.0 m) from the ground due to insufficient hoist rope on the crane.

Additional test sets were included to examine other aspects of PWD performance, such as using maximum device sensitivity for the time a crane is in transit in order to alert the operator and crew that power lines are somewhere in or near an unfamiliar work area as the crane approaches. Ideally, this warning would be well before the crane is in close proximity to the lines. Test set 8 placed the crane in a travel configuration, approaching a single set of horizontal power lines with PWDs set at maximum sensitivity to determine the greatest distance at which the lines could be detected. Test set 9 had the crane, with PWDs again set at maximum sensitivity, approach two sets of horizontal lines intersecting at 90° to determine whether the alarm distance was influenced by the additional set of power lines.

Test set 10 examined the effect of having the crane chassis tied directly to the power system's grounded neutral while operating from a stationary position. Connecting an equipment chassis to ground is sometimes suggested as a means to reduce the chance of a dangerous chassis voltage in the event of an accidental power line contact or to reduce induced charges on a piece of equipment.

RESULTS

The following sections present performance data for the two PWDs tested. Each section lists test and sensitivity adjustment details, provides a diagram of the test layout and power line configuration, and presents a table showing the distances from the overhead power lines at which the PWDs alarmed for specific conditions (or notes no alarm activation). The alarm distances, as noted earlier, are true distances taken from the 3-D models created for each test.

Test Set 1

Test set 1 used one set of horizontal 13-kV lines. Sensitivity was adjusted once prior to ABC phase sequence tests (A-phase nearest crane) and once prior to CBA phase sequence tests (C-phase nearest crane). Sensitivity was adjusted with the crane parked in the test position on its outriggers, the boom extended to approximately 43 ft (13.1 m) (62% of full extension), the boom tip elevated to approximately phase conductor level, and the boom swung to a position 20 ft (6.1 m) from the closest conductor.[11] Test layout and power line configuration are shown in Figure 9. Table 2 details performance of the PWDs for test set 1.

---

[8]These extensions give boom lengths of 28 ft (8.5 m), 48.7 ft (14.8 m), and 69.4 ft (21.2 m), respectively, from base pivot pin to point sheave pin.

[9]Sixty-one degrees was the maximum elevation angle (at minimum boom length) at which the load hook was not over the crane chassis.

[10]With the boom at "minimum boom length," it could not be positioned with the "boom tip at power line height" for any power line configuration used. Therefore, this combination was not possible.

[11]As described earlier, the distance listed for the sensitivity adjustments is the "nominal" distance from the power lines, which was determined by using the distance reference grid painted on the ground at the test site. There are slight variations in the sensitivity adjustment boom lengths listed because the boom positions were approximated in the field and then later measured precisely from the survey data-based computer models.

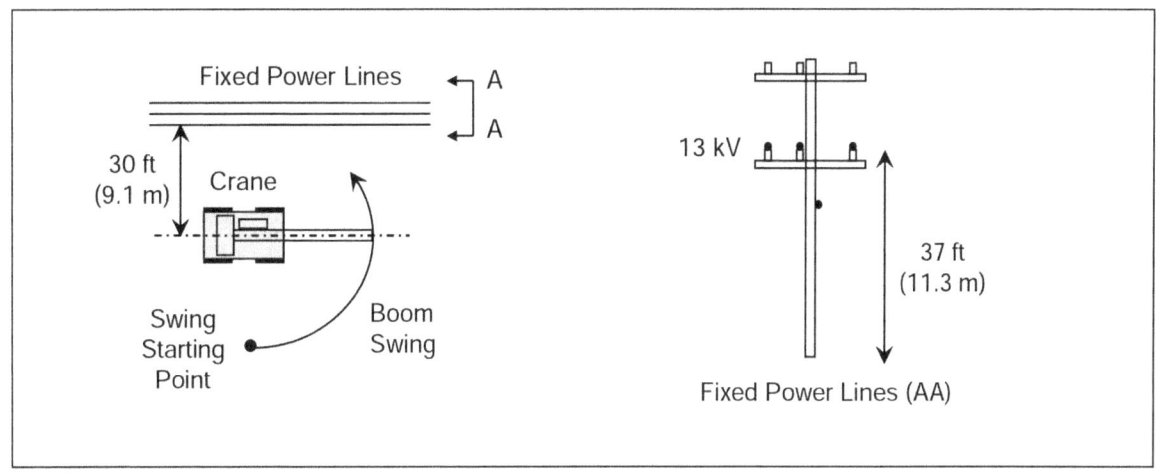

Figure 9.—Test set 1 layout and overhead power line configuration.

Table 2.—Alarm distance results for test set 1
(Distance between boom or hoist ropes and closest phase conductor in ft (m))

Phase sequence: horizontal ABC with A-phase nearest crane							
SIGALARM Model 210				ASE Model 2100			
Boom position	Minimum extension	50% extension	Maximum extension		Minimum extension	50% extension	Maximum extension
0° elevation	No alarm	No alarm	No alarm		No alarm	No alarm	No alarm
Boom tip at power line level	([1])	27.6 (8.4)	36.1 (11.0)		([1])	27.6 (8.4)	51.8 (15.8)
61° elevation	No alarm	19.4 (5.9)	No alarm	61° Elevation	No alarm	29.3 (8.9)	48.6 (14.8)
Phase sequence: horizontal CBA with C-phase nearest crane							
SIGALARM Model 210				ASE Model 2100			
Boom position	Minimum extension	50% extension	Maximum extension		Minimum extension	50% extension	Maximum extension
0° elevation	No alarm	No alarm	No alarm		No alarm	No alarm	No alarm
Boom tip at power line level	([1])	23.4 (7.1)	38.6 (11.8)		([1])	29.0 (8.8)	53.4 (16.3)
61° elevation	No Alarm	No Alarm	No Alarm	61° elevation	No alarm	28.7 (8.7)	52.9 (16.1)

[1]The boom tip could not be raised to power line level with the boom at minimum extension length for any of the power line configurations used; therefore, this combination of extension and elevation angle was not possible.

Test Set 2

Test set 2 used one set of vertical 25-kV lines. Sensitivity was adjusted once prior to ABC phase sequence tests (A-phase at top) and once prior to CBA phase sequence tests (C-phase at top). Sensitivity was adjusted with the crane parked in the test position on its outriggers, the boom extended to approximately 57 ft (17.4 m) (82% of full extension), the boom tip elevated to approximately top phase conductor level, and the boom swung to a position 20 ft (6.1 m) from the top conductor. Test layout and power line configuration are shown in Figure 10. Table 3 details performance of the PWDs for test set 2.

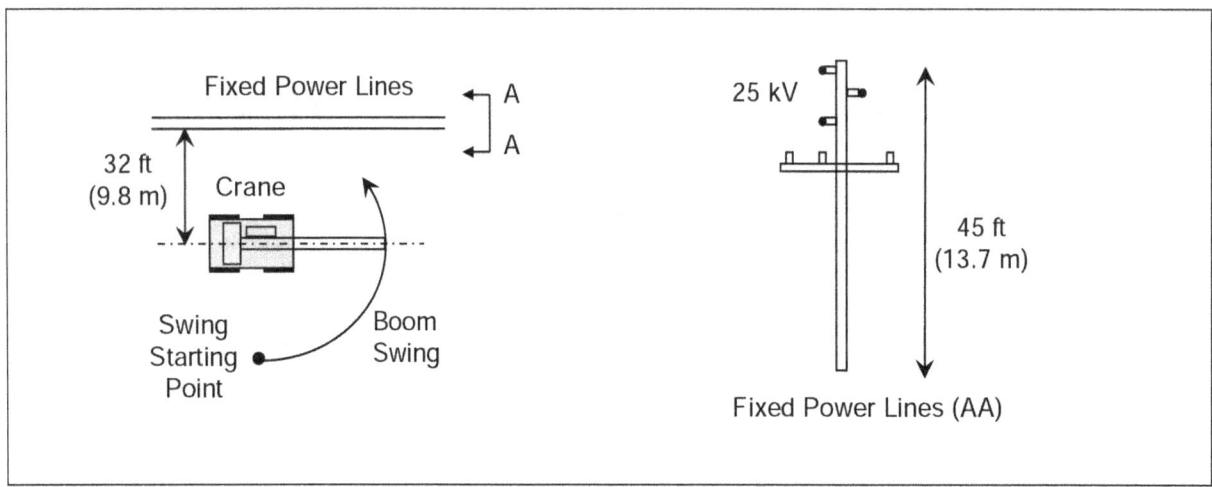

Figure 10.—Test set 2 layout and overhead power line configuration.

Table 3.—Alarm distance results for test set 2
(Distance between boom or hoist ropes and closest phase conductor in ft (m))

Phase sequence: vertical ABC with A-phase at top (25 kV)							
SIGALARM Model 210				ASE Model 2100			
Boom position	Minimum extension	50% extension	Maximum extension		Minimum extension	50% extension	Maximum extension
0° elevation	No alarm	No alarm	No alarm		No alarm	No alarm	No alarm
Boom tip at power line level	(^{1})	(^{2})	31.7 (9.7)		(^{1})	(^{2})	31.7 (9.7)
61° elevation	No alarm	12.1 (3.7)	21.8 (6.6)	61° elevation	No alarm	12.1 (3.7)	33.4 (10.2)
Phase sequence: vertical CBA with C-phase at top (25 kV)							
SIGALARM Model 210				ASE Model 2100			
Boom position	Minimum extension	50% extension	Maximum extension		Minimum extension	50% extension	Maximum extension
0° elevation	No alarm	No alarm	No alarm		No alarm	No alarm	No alarm
Boom tip at power line level	(^{1})	(^{2})	32.1 (9.8)		(^{1})	(^{2})	32.1 (9.8)
61° elevation	No alarm	8.1 (2.5)	28.3 (8.6)	61° elevation	No alarm	No alarm	28.3 (8.6)

[1] The boom tip could not be raised to power line level with the boom at minimum extension length for any of the power line configurations used; therefore, this combination of extension and elevation angle was not possible.
[2] These tests were omitted in error.

Test Set 3

Test set 3 used two sets of horizontal 13-kV lines supported on the same poles. Sensitivity was adjusted once prior to tests using ABC phase sequence on both sets of lines (A-phase nearest crane for both sets) and once prior to tests using CBA phase sequence on the upper set of lines (C-phase nearest crane for upper set). Sensitivity was adjusted with the crane parked in the test position on its outriggers, the boom extended to approximately 44 ft (13.4 m) (63% of full extension), the boom tip elevated to approximately phase conductor level for the lower set of lines, and the boom swung to a position 20 ft (6.1 m) from the closest conductor. Test layout and power line configuration are shown in Figure 11. Table 4 details performance of the PWDs for test set 3.

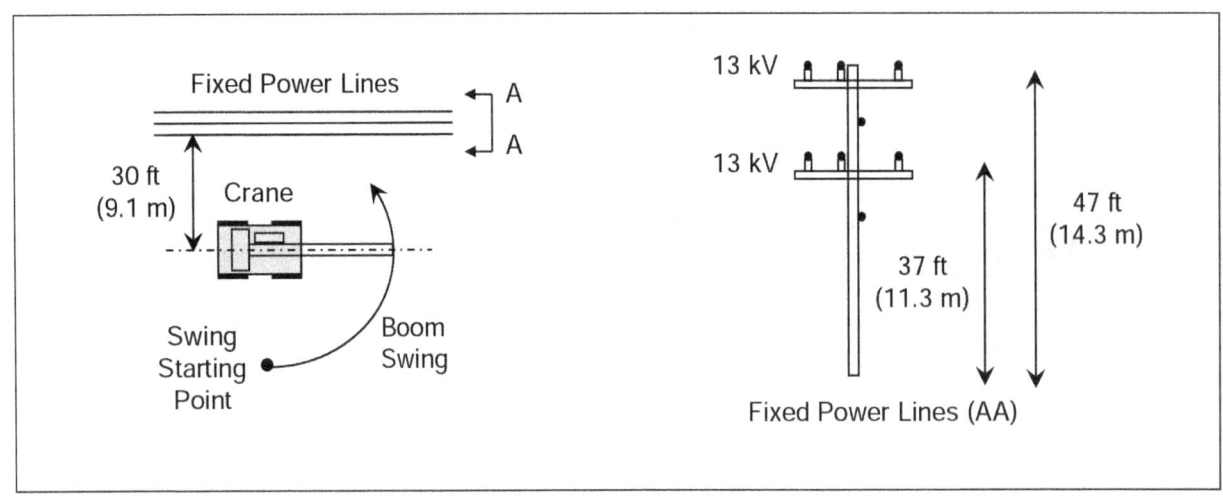

Figure 11.—Test set 3 layout and overhead power line configuration.

Table 4.—Alarm distance results for test set 3
(Distance between boom or hoist ropes and closest phase conductor in ft (m))

Upper set of power lines – phase sequence: horizontal ABC with A-phase nearest crane Lower set of power lines – phase sequence: horizontal ABC with A-phase nearest crane							
SIGALARM Model 210				ASE Model 2100			
Boom position	Minimum extension	50% extension	Maximum extension		Minimum extension	50% extension	Maximum extension
0° elevation	No alarm	No alarm	No alarm		No alarm	No alarm	No alarm
Boom tip at power line level	(^{1})	27.2 (8.3)	45.0 (13.7)		(^{1})	^{2}35.0 (10.7)	60.9 (18.6)
61° elevation	No alarm	29.6 (9.0)	44.9 (13.7)	61° elevation	No alarm	37.0 (11.3)	62.3 (19.0)
Upper set of power lines – phase sequence: horizontal CBA with C-phase nearest crane Lower set of power lines – phase sequence: horizontal ABC with A-phase nearest crane							
SIGALARM Model 210				ASE Model 2100			
Boom position	Minimum extension	50% extension	Maximum extension		Minimum extension	50% extension	Maximum extension
0° elevation	No alarm	No alarm	No alarm		No alarm	No alarm	No alarm
Boom tip at power line level	(^{1})	^{2}24.0 (7.3)	31.9 (9.7)		(^{1})	^{2}24.0 (7.3)	38.4 (11.7)
61° elevation	No alarm	15.0 (4.6)	No alarm	61° elevation	No alarm	22.3 (6.8)	35.5 (10.8)

^{1}The boom tip could not be raised to power line level with the boom at minimum extension length for any of the power line configurations used; therefore, this combination of extension and elevation angle was not possible.

^{2}Value is the observed distance due to invalid survey data for the boom position.

Test Set 4

Test set 4 used two sets of horizontal power lines supported on the same poles. The upper set was at 25 kV; the lower set was at 13 kV. Sensitivity was adjusted once prior to tests using ABC phase sequence on both sets of lines (A-phase nearest crane for both sets) and once prior to tests using CBA phase sequence on the upper set of lines (C-phase nearest crane for upper set). Sensitivity was adjusted with the crane parked in the test position on its outriggers, the boom extended to approximately 43 ft (13.1 m) (62% of full extension), the boom tip elevated to approximately phase conductor level for the lower set of lines, and the boom swung to a position 20 ft (5.8 m) from the closest conductor. Test layout and power line configuration are shown in Figure 12. Table 5 details performance of the PWDs for test set 4.

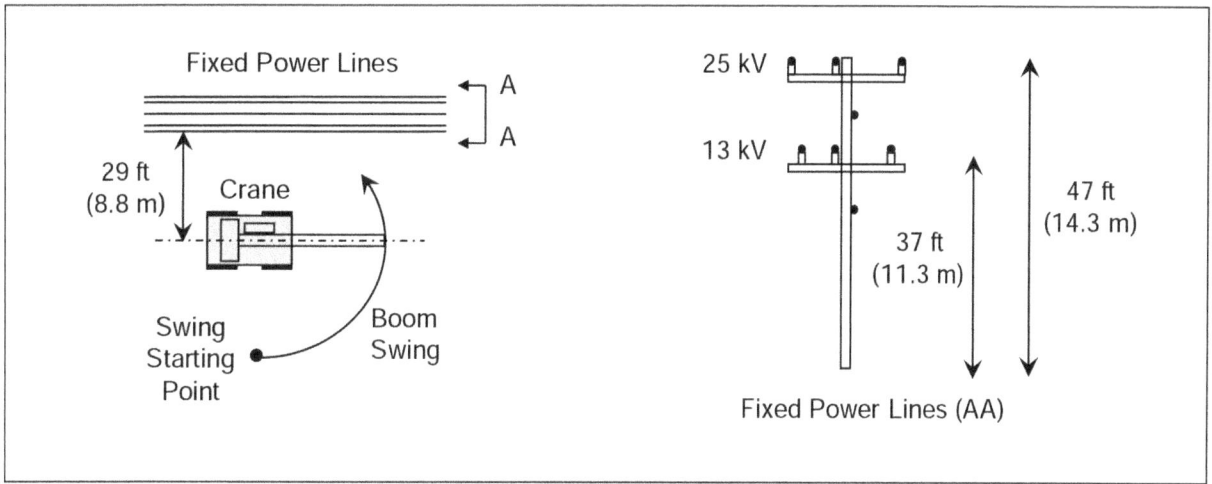

Figure 12.—Test set 4 layout and overhead power line configuration.

Table 5.—Alarm distance results for test set 4
(Distance between boom or hoist ropes and closest phase conductor in ft (m))

Upper set of power lines – phase sequence: horizontal ABC with A-phase nearest crane (25 kV) Lower set of power lines – phase sequence: horizontal ABC with A-phase nearest crane							
	SIGALARM Model 210				ASE Model 2100		
Boom position	Minimum extension	50% extension	Maximum extension		Minimum extension	50% extension	Maximum extension
0° elevation	No alarm	No alarm	No alarm		No alarm	No alarm	No alarm
Boom tip at power line level	([1])	22.8 (6.9)	37.9 (11.6)		([1])	29.3 (8.9)	57.7 (17.6)
61° elevation	No alarm	24.6 (7.5)	[2]43.0 (13.1)	61° elevation	No alarm	34.5 (10.5)	61.6 (18.8)
Upper set of power lines – phase sequence: horizontal CBA with C-phase nearest crane (25 kV) Lower set of power lines – phase sequence: horizontal ABC with A-phase nearest crane							
	SIGALARM Model 210				ASE Model 2100		
Boom position	Minimum extension	50% extension	Maximum extension		Minimum extension	50% extension	Maximum extension
0° elevation	No alarm	No alarm	No alarm		No alarm	No alarm	No alarm
Boom tip at power line level	([1])	21.1 (6.4)	33.6 (10.2)		([1])	29.7 (9.1)	57.3 (17.5)
61° elevation	No alarm	25.3 (7.7)	40.8 (12.4)	61° elevation	No alarm	38.7 (11.8)	62.0 (18.9)

[1]The boom tip could not be raised to power line level with the boom at minimum extension length for any of the power line configurations used; therefore, this combination of extension and elevation angle was not possible.

[2]Value is the observed distance due to invalid survey data for the boom position.

Test Set 5

Test set 5 used two sets of power lines supported on the same poles. The upper set was a vertical configuration at 25 kV; the lower set was a horizontal configuration at 13 kV. Sensitivity was adjusted once prior to tests using ABC phase sequence on both sets of lines (A-phase nearest crane for horizontal set and at top for vertical set) and once prior to tests using CBA phase sequence on the upper set of lines (C-phase at top), with the lower set remaining ABC. Sensitivity was adjusted with the crane parked in the test position on its outriggers, the boom extended to approximately 44 ft (13.4 m) (63% of full extension), the boom tip elevated to approximately phase conductor level for the lower set of lines, and the boom swung to a position 20 ft (6.1 m) from the closest conductor. Test layout and power line configuration are shown in Figure 13. Table 6 details performance of the PWDs for test set 5.

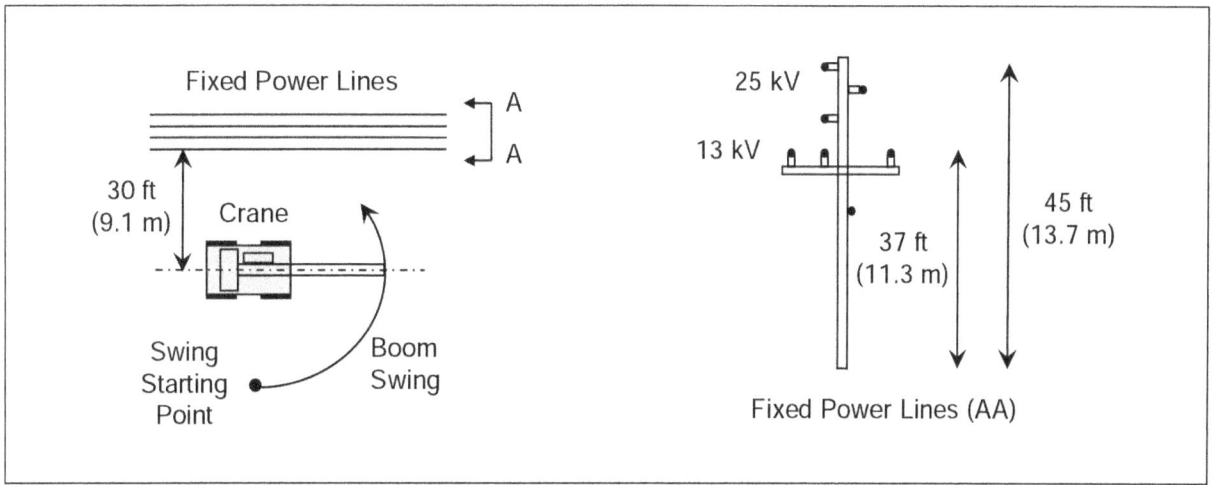

Figure 13.—Test set 5 layout and overhead power line configuration.

Table 6.—Alarm distance results for test set 5
(Distance between boom or hoist ropes and closest phase conductor in ft (m))

| | Upper set of power lines – phase sequence: vertical ABC with A-phase at top (25 kV)
Lower set of power lines – phase sequence: horizontal ABC with A-phase nearest crane |

	SIGALARM Model 210				ASE Model 2100		
Boom position	Minimum extension	50% extension	Maximum extension		Minimum extension	50% extension	Maximum extension
0° elevation	No alarm	No alarm	34.2 (10.4)		No alarm	No alarm	No alarm
Boom tip at power line level	(^{1})	49.4 (15.1)	74.3 (22.6)		(^{1})	41.8 (12.7)	78.6 (24.0)
61° elevation	No alarm	^{2}52.2 (15.9)	^{2}65.0 (19.8)	61° elevation	No alarm	46.0 (14.0)	^{2}65.0 (19.8)

Upper set of power lines – phase sequence: vertical CBA with C-phase at top (25 kV)
Lower set of power lines – phase sequence: horizontal ABC with A-phase nearest crane

	SIGALARM Model 210				ASE Model 2100		
Boom position	Minimum extension	50% extension	Maximum extension		Minimum extension	50% extension	Maximum extension
0° elevation	No alarm	35.0 (10.7)	41.2 (12.6)		No alarm	No alarm	29.0 (8.8)
Boom tip at power line level	(^{1})	20.5 (6.2)	31.1 (9.5)		(^{1})	20.5 (6.2)	31.1 (9.5)
61° elevation	No alarm	31.9 (9.7)	^{2}63.6 (19.4)	61° elevation	No alarm	11.0 (3.4)	51.1 (15.6)

^{1}The boom tip could not be raised to power line level with the boom at minimum extension length for any of the power line configurations used; therefore, this combination of extension and elevation angle was not possible.
^{2}Alarmed at swing starting point.

Test Set 6

Test set 6 used two separate sets of power lines intersecting at 90°, with the centerline of the crane bisecting the angle between the power lines. Both sets of lines were in a horizontal configuration operating at 13 kV. The lines to the crane's right (from the operator's perspective) were on movable poles; those to the left were on fixed poles. Sensitivity was adjusted once prior to tests using ABC phase sequence on both sets of lines (A-phase nearest crane for both sets) and once prior to tests using ABC phase sequence on the fixed set and CBA on the movable set. Sensitivity was adjusted with the crane parked in the test position on its outriggers, the boom extended to 42 ft (12.8 m) (61% of full extension), the boom tip elevated to approximately phase conductor level for the lines on the left (fixed lines), and the boom swung left to a position 20 ft (6.1 m) from the closest conductor. Test layout and power line configuration are shown in Figure 14. Table 7 details performance of the PWDs for test set 6.

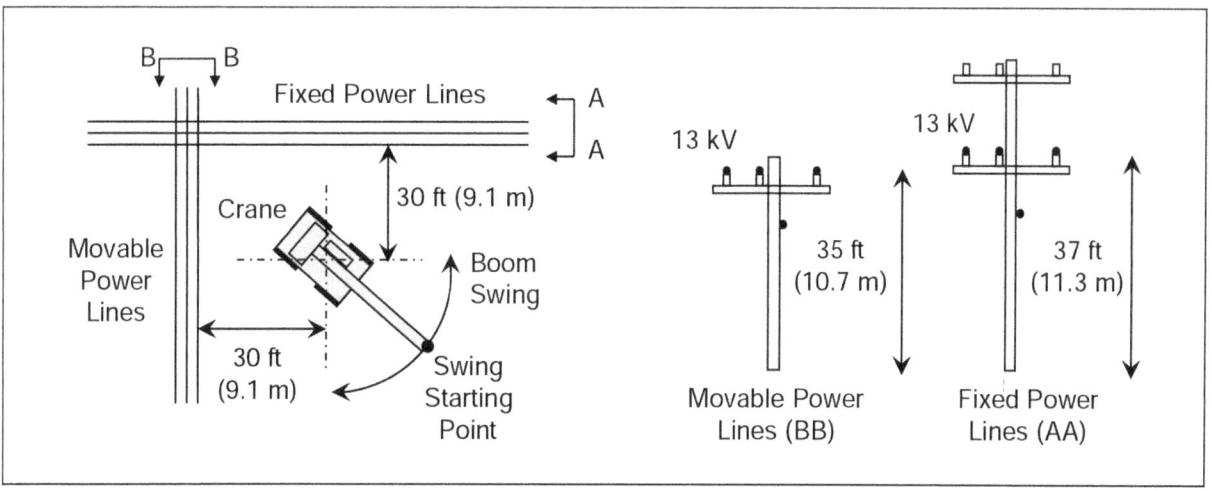

Figure 14.—Test set 6 layout and overhead power line configuration.

Table 7.—Alarm distance results for test set 6
(Distance between boom or hoist ropes and closest phase conductor in ft (m))

Fixed power lines – phase sequence: horizontal ABC with A-phase nearest crane (left swing) Movable power lines – phase sequence: horizontal ABC with A-phase nearest crane (right swing)							
	SIGALARM Model 210				ASE Model 2100		
Boom position	Minimum extension	50% extension	Maximum extension		Minimum extension	50% extension	Maximum extension
0° elevation	Left swing No alarm	Left swing No alarm	Left swing No alarm		Left swing No alarm	Left swing No alarm	Left swing No alarm
	Right swing No alarm	Right swing No alarm	Right swing No alarm		Right swing No alarm	Right swing No alarm	Right swing No alarm
Boom tip at power line level	([1])	Left swing 29.3 (8.9)	Left swing 42.4 (12.9)	([1])		Left swing 29.3 (8.9)	Left swing [2]70.2 (21.4)
		Right swing 18.0 (5.5)	Right swing 33.6 (10.2)			Right swing 35.3 (10.8)	Right swing [2]72.6 (22.1)
61° elevation	Left swing No alarm	Left swing 36.6 (11.2)	Left swing [2]53.4 (16.3)	61° elevation	Left swing No alarm	Left swing 36.6 (11.2)	Left swing [2]53.4 (16.3)
	Right swing No alarm	Right swing No alarm	Right swing [2]53.4 (16.3)		Right swing No alarm	Right swing 39.0 (11.9)	Right swing [2]53.4 (16.3)

Fixed power lines – phase sequence: horizontal CBA with C-phase nearest crane (left swing) Movable power lines – phase sequence: horizontal ABC with A-phase nearest crane (right swing)							
	SIGALARM Model 210				ASE Model 2100		
Boom position	Minimum extension	50% extension	Maximum extension		Minimum extension	50% extension	Maximum extension
0° elevation	Left swing No alarm	Left swing No alarm	Left swing No alarm		Left swing No alarm	Left swing No alarm	Left swing No alarm
	Right swing No alarm	Right swing No alarm	Right swing 29.8 (9.1)		Right swing No alarm	Right swing No alarm	Right swing 29.8 (9.1)
Boom tip at power line level	([1])	Left swing 29.1 (8.9)	Left swing 38.7 (11.8)	([1])		Left swing 29.1 (8.9)	Left swing 48.5 (14.8)
		Right swing 14.4 (4.4)	Right swing 84.0 (25.6)			Right swing 32.0 (9.8)	Right swing 51.2 (15.6)
61° elevation	Left swing No alarm	Left swing No alarm	Left swing No alarm	61° elevation	Left swing No alarm	Left swing 26.2 (8.0)	Left swing 41.9 (12.8)
	Right swing No alarm	Right swing No alarm	Right swing No alarm		Right swing No alarm	Right swing 27.8 (8.5)	Right swing 40.8 (12.4)

[1]The boom tip could not be raised to power line level with the boom at minimum extension length for any of the power line configurations used; therefore, this combination of extension and elevation angle was not possible.
[2]Alarmed at swing starting point.

Test Set 7

Test set 7 used two separate sets of power lines intersecting at 90°, with the centerline of the crane bisecting the angle between the power lines. Both sets of lines were in a horizontal configuration. The lines to the crane's right (from the operator's perspective) were on movable poles and operating at 13 kV; those to the left were on fixed poles and operating at 25 kV. Sensitivity was adjusted once prior to tests using ABC phase sequence on both sets of lines (A-phase nearest crane for both sets) and once prior to tests using ABC phase sequence on the fixed set and CBA on the movable set (C-phase nearest crane). Sensitivity was adjusted with the crane parked in the test position on its outriggers, the boom extended to approximately 44 ft (13.4 m) (63% of full extension), the boom tip elevated to approximately phase conductor level for the lines on the right (13-kV movable lines), and the boom swung right to a position 20 ft (6.1 m) from the closest conductor. Test layout and power line configuration are shown in Figure 15. Table 8 details performance of the PWDs for test set 7.

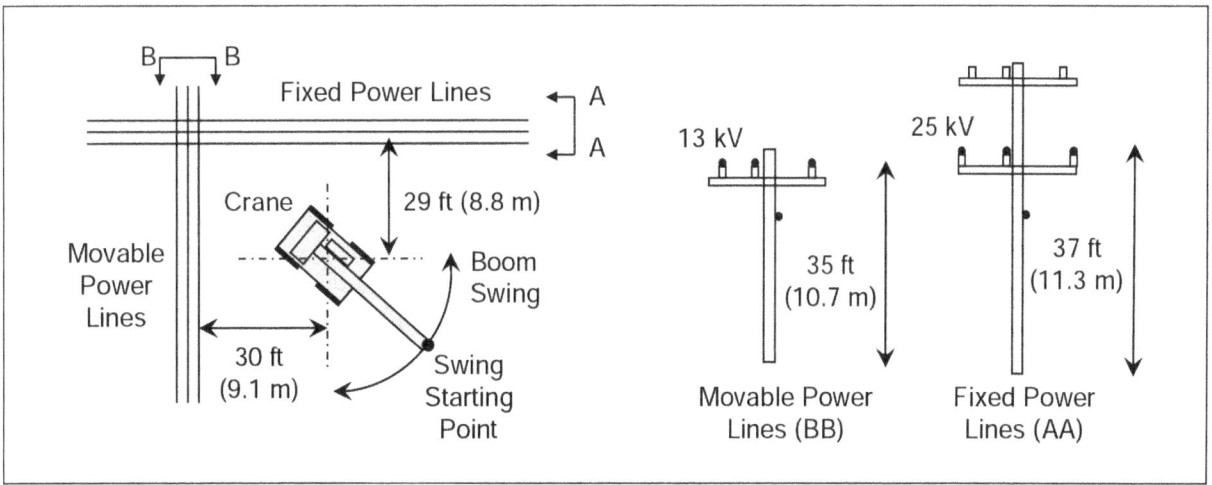

Figure 15.—Test set 7 layout and overhead power line configuration.

Table 8.—Alarm distance results for test set 7
(Distance between boom or hoist ropes and closest phase conductor in ft (m))

Fixed power lines – phase sequence: horizontal ABC with A-phase nearest crane (left swing) (25 kV)
Movable power lines – phase sequence: horizontal ABC with A-phase nearest crane (right swing)

Boom position	SIGALARM Model 210			ASE Model 2100		
	Minimum extension	50% extension	Maximum extension	Minimum extension	50% extension	Maximum extension
0° elevation	*Left swing* No alarm	*Left swing* 29.7 (9.1)	*Left swing* 50.6 (15.4)	*Left swing* No alarm	*Left swing* No alarm	*Left swing* No alarm
	Right swing No alarm	*Right swing* No alarm	*Right swing* No alarm	*Right swing* No alarm	*Right swing* No alarm	*Right swing* No alarm
Boom tip at power line level	([1])	*Left swing* [2]54.2 (16.5)	*Left swing* [2]70.9 (21.6)	([1])	*Left swing* [2]54.2 (16.5)	*Left swing* [2]70.9 (21.6)
		Right swing [2]56.9 (17.3)	*Right swing* [2]73.8 (22.5)		*Right swing* [2]56.9 (17.3)	*Right swing* [2]73.8 (22.5)
61° elevation	*Left swing* 26.5 (8.1)	*Left swing* [2]44.7 (13.6)	*Left swing* [2]52.8 (16.1)	*Left swing* No alarm	*Left swing* [2]44.7 (13.6)	*Left swing* [2]52.8 (16.1)
	Right swing No alarm	*Right swing* [2]45.3 (13.8)	*Right swing* [2]55.1 (16.8)	*Right swing* No alarm	*Right swing* [2]45.3 (13.8)	*Right swing* [2]55.1 (16.8)

Fixed power lines – phase sequence: horizontal ABC with A-phase nearest crane (left swing) (25 kV)
Movable power lines – phase sequence: horizontal CBA with C-phase nearest crane (right swing)

Boom position	SIGALARM Model 210			ASE Model 2100		
	Minimum extension	50% extension	Maximum extension	Minimum extension	50% extension	Maximum extension
0° elevation	*Left swing* 43.0 (13.1)	*Left swing* [2]65.2 (19.9)	*Left swing* [2]80.7 (24.6)	*Left swing* No alarm	*Left swing* 41.9 (12.8)	*Left swing* 63.8 (19.5)
	Right swing No alarm	*Right swing* [2]65.2 (19.9)	*Right swing* [2]79.7 (24.3)	*Right swing* No alarm	*Right swing* No alarm	*Right swing* No alarm
Boom tip at power line level	([1])	*Left swing* [2]54.5 (16.6)	*Left swing* [2]71.6 (21.8)	([1])	*Left swing* [2]54.2 (16.5)	*Left swing* [2]71.6 (21.8)
		Right swing [2]56.2 (17.1)	*Right swing* [2]73.3 (22.3)		*Right swing* [2]56.2 (17.1)	*Right swing* [2]73.3 (22.3)
61° elevation	*Left swing* [2]36.8 (11.2)	*Left swing* [2]43.9 (13.4)	*Left swing* [2]52.0 (15.9)	*Left swing* 24.0 (7.3)	*Left swing* [2]43.9 (13.4)	*Left swing* [2]52.0 (15.9)
	Right swing No alarm	*Right swing* [2]44.6 (13.6)	*Right swing* [2]52.9 (16.1)	*Right swing* No alarm	*Right swing* [2]44.6 (13.6)	*Right swing* [2]52.9 (16.1)

[1]The boom tip could not be raised to power line level with the boom at minimum extension length for any of the power line configurations used; therefore, this combination of extension and elevation angle was not possible.
[2]Alarmed at swing starting point.

For several specific cases in test set 7, while the boom was swinging toward the lines with the PWDs in an alarm state, the alarms would cease and then resume closer to the lines. For the SIGALARM Model 210, with CBA phase sequence on the movable lines, the right swing tests combining 0° elevation with 50% and 100% extensions had a cessation of the alarm during the swing and resumption at 27.8 ft (8.5 m) and 34.2 ft (10.4), respectively. This cessation also occurred for the ASE Model 2100 for right swing tests combining the "boom tip at line height" elevation with 50% and 100% extensions, with the alarm resuming at 23.9 ft (7.3 m) and 33.3 ft (10.1 m), respectively. This behavior was noted only in these cases, but not all tests were continued past the point of alarms to check for alarm cessation.

Test Set 8

Test set 8 used one set of horizontal 13-kV lines with ABC phase sequence (A-phase nearest crane). Both PWDs were adjusted to maximum sensitivity (alarm activation at greatest distance from power lines). With the boom at minimum extension and elevation and centered over the front of the crane (travel configuration), the crane was driven toward the power lines at an approach angle of 90°, 45° from the right, and 45° from the left. At the point of alarm activation, the crane was stopped and set up on outriggers, and the boom was extended to maximum length and rotated to the swing starting point. Tests were then conducted by swinging the boom toward the power lines. Test layout and power line configuration are shown in Figure 16.

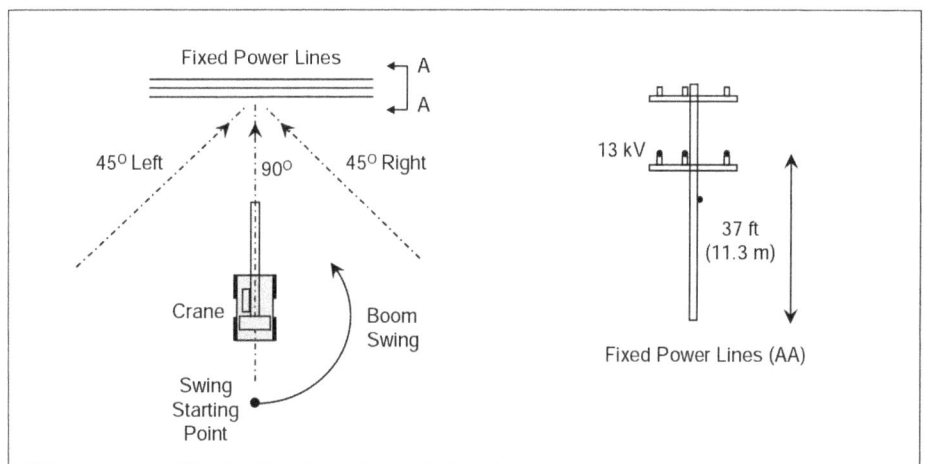

Figure 16.—Test set 8 layout and overhead power line configuration.

With the crane traveling toward the power lines, the SIGALARM Model 210, adjusted to maximum sensitivity, alarmed at 82.0 ft (25.0 m), 83.4 ft (25.4 m), and 75.1 ft (22.9 m) from the power lines for 90°, 45° right, and 45° left approaches, respectively (measured perpendicular to the power lines). When the crane was set up at the 90° approach alarm point and the fully extended boom was swung left toward the lines at 0° elevation, boom tip at line height, and 61° elevation, the Model 210 alarmed at 147.5 ft (45.0 m), 165.6 ft (50.5 m), and 124.0 ft (38.1 m), respectively. Based on these results, boom swing tests were omitted for the 45° approach alarm points.

With the crane traveling toward the power lines, the ASE Model 2100 at maximum sensitivity did not alarm while approaching the power lines. The tests were terminated when the crane boom was directly under the lines.

Test Set 9

Test set 9 used two sets of horizontal 13-kV lines with ABC phase sequence (A-phase nearest crane) intersecting at 90°. Both PWDs were adjusted to maximum sensitivity (alarm activation at greatest distance from power lines). With the boom at minimum extension and elevation and centered over the front of the crane, the crane was driven toward the power lines on a path bisecting the angle between the lines. At the point of alarm activation, the crane was stopped and set up on outriggers, and the boom was extended to maximum length and rotated to the swing starting point. Tests were then conducted by swinging the boom toward the power lines. Test layout and power line configuration are shown in Figure 17.

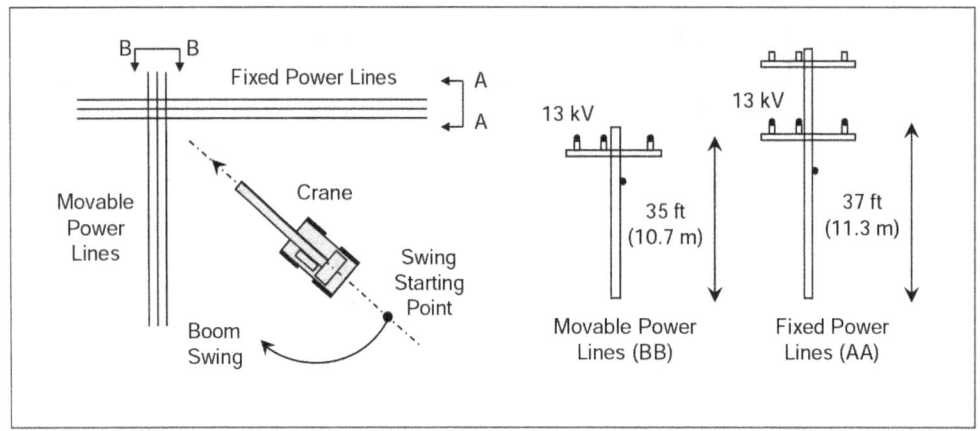

Figure 17.—Test set 9 layout and overhead power line configuration.

With the crane traveling toward the power lines, the SIGALARM Model 210, adjusted to maximum sensitivity, alarmed at 88 ft (26.8 m) from both sets of power lines (observed distance measured perpendicular to both sets of power lines). When the crane was set up at this location and the fully extended boom was swung right toward the portable power lines, the Model 210 alarmed at 154.2 ft (47.0 m) and 136.2 ft (41.5 m) for the boom tip at power line height and 61° elevation, respectively. No data were obtained for 0° boom elevation.

With the crane traveling toward the power lines, the ASE Model 2100, adjusted to maximum sensitivity, alarmed at 20 ft (6.1 m) from the power lines. When the crane was set up at this location and the fully extended boom was swung right toward the portable power lines, the Model 2100 alarmed at 70.1 ft (21.4 m), 62.3 ft (19.0 m), and 52.7 ft (16.1 m) for 0° elevation, boom tip at power line height, and 61° elevation, respectively.

Test Set 10

Connecting an equipment chassis to ground is sometimes suggested as a means to reduce the chance of a dangerous chassis voltage in the event of an accidental power line contact or to avoid an induced charge on a piece of equipment. Test set 10 examined the effect on PWD performance of connecting the crane chassis directly to the grounded power system neutral. These tests used one set of horizontal 13-kV lines with ABC phase sequence (A-phase nearest crane). Sensitivity was adjusted once prior to testing with the crane parked in the test position on its outriggers, the boom extended to 46 ft (14.0 m) (66% of full extension), the boom tip elevated to approximately phase conductor level, and the boom swung to a position 20 ft (6.1 m) from the closest conductor. The same eight boom positions were used as in test sets 1 through 7. One complete series was done with the crane chassis tied directly to the grounded power system neutral, then those tests were duplicated with the chassis disconnected from the system neutral (initial sensitivity adjustments were left in place for both series of tests).

For both PWDs, the tests showed no significant difference in the alarm distances between the series with and without the crane chassis tied directly to the power system ground.

SUMMARY OF PROXIMITY WARNING DEVICE PERFORMANCE

A review of the foregoing test results clearly illustrates that many, often interrelated, factors affect performance of PWDs. One of the most important factors is sensitivity adjustment. Both PWD manufacturers cooperating in this research adjusted the sensitivity of their devices in the same manner for each set of tests. This adjustment used a relatively consistent boom position and configuration and a nominal distance of approximately 20 ft (6.1 m) from the power lines. This adjustment was performed only twice for each of test sets 1 through 7, as detailed earlier. While this approach did not optimize PWD performance for any given test, it had the advantage of allowing comparisons of the effects of changes in power line and boom configurations. Most of the performance assessments presented in this section highlight and rank the effects of these factors on the accuracy and reliability of the two PWDs tested.

Power Line Configurations

Test sets 1 through 7 examined the effects of differences in power line configurations on PWD performance and included 284 individual tests.[12] Of these 284 tests, 108 had a boom elevation of 0° (boom approximately level with the ground), which allowed the boom to be swung below the power lines without coming within 20 ft (6.1 m) (true distance) of any phase conductor (20 ft (6.1 m) was the nominal distance used for PWD sensitivity adjustments). As discussed earlier, operating with the boom directly above or below overhead power lines can be very hazardous and should normally be avoided. For these 108 tests, however, the horizontal boom passing below the power lines without triggering an alarm is considered acceptable PWD operation, since the boom was never closer than 20 ft (6.1 m) (true distance) from an energized conductor. Of these 108 cases, 15 caused a PWD alarm, which by the criteria just described constitutes an "early alarm." The remaining 176 tests (those with the boom elevated such that the tip is at power line height or elevated to 61°) should have all resulted in PWD alarms, ideally at approximately 20 ft (6.1 m) from the closest phase conductor. Of these 176 tests, 148 produced alarms. The 28 remaining tests that did not cause an alarm were stopped when the boom was 10 ft (3.0 m) from the power lines. Of the 148 alarm activations, 140 were early alarms (more than 20 ft (6.1 m) from power lines), and 8 were "late alarms" (less than 20 ft (6.1 m) and more than 10 ft (3.0 m) from the lines).

Table 9 summarizes PWD alarm performance based on different power line configurations for test sets 1 through 7. The table lists the number of no alarm, early alarm, and late alarm cases for each test set, as well as the average of alarm distance deviations from 20 ft (6.1 m) (magnitudes) for the tests that caused alarm activations.

---

[12] The number 284 is based on counting the same test run for each PWD as two tests.

Table 9.—Summary of PWD alarm performance based on power line configuration

Test set No. and power line configuration		SIGALARM Model 210		ASE Model 2100	
		Alarm record: (a) No alarm[1] (b) Late alarm[1] (c) Early alarm[2]	Alarm distance deviation,[3] ft (m)	Alarm record: (a) No alarm[1] (b) Late alarm[1] (c) Early alarm[2]	Alarm distance deviation,[3] ft (m)
1	One horizontal line set, 13 kV	(a) 5 of 10 (b) 1 of 10 (c) 4 of 16	9.4 (2.9)	(a) 2 of 10 (b) 0 of 10 (c) 8 of 16	20.3 (6.2)
2	One vertical line set, 25 kV	(a) 2 of 8 (b) 2 of 8 (c) 4 of 14	9.0 (2.7)	(a) 3 of 8 (b) 1 of 8 (c) 4 of 14	10.6 (3.2)
3	Two horizontal line sets on same poles, 13 kV/13 kV	(a) 3 of 10 (b) 1 of 10 (c) 6 of 16	12.6 (3.8)	(a) 2 of 10 (b) 0 of 10 (c) 8 of 16	19.4 (5.9)
4	Two horizontal line sets on same poles, 25 kV/13 kV	(a) 2 of 10 (b) 0 of 10 (c) 8 of 16	13.5 (4.1)	(a) 2 of 10 (b) 0 of 10 (c) 8 of 16	27.5 (8.4)
5	Two line sets, one vertical over one horizontal on same poles, 25 kV/13 kV	(a) 2 of 10 (b) 0 of 10 (c) 11 of 16	27.5 (8.4)	(a) 2 of 10 (b) 1 of 10 (c) 8 of 16	23.9 (7.3)
6	Two horizontal line sets crossing at 90°, 13 kV/13 kV *Left swing*	(a) 4 of 10 (b) 0 of 10 (c) 6 of 16	18.2 (5.5)	(a) 2 of 10 (b) 0 of 10 (c) 8 of 16	21.9 (6.7)
6	Two horizontal line sets crossing at 90°, 13 kV/13 kV *Right swing*	(a) 5 of 10 (b) 2 of 10 (c) 4 of 16	23.8 (7.3)	(a) 2 of 10 (b) 0 of 10 (c) 9 of 16	22.7 (6.9)
7	Two horizontal line sets crossing at 90°, 25 kV/13 kV *Left swing toward 25-kV lines*	(a) 0 of 10 (b) 0 of 10 (c) 15 of 16	35.6 (10.9)	(a) 1 of 10 (b) 0 of 10 (c) 11 of 16	34.5 (10.5)
7	Two horizontal line sets crossing at 90°, 25 kV/13 kV *Right swing toward 13-kV lines*	(a) 2 of 10 (b) 0 of 10 (c) 10 of 16	43.2 (13.2)	(a) 2 of 10 (b) 0 of 10 (c) 8 of 16	37.3 (11.4)

[1] The denominator in this entry refers to tests in the category listed, where an alarm should have occurred at approximately 20 ft (6.1 m) from the power lines.

[2] The denominator in this entry refers to all tests in the category listed.

[3] Deviation refers to the average of the magnitudes for (true alarm distance – 20 ft (6.1 m)), where 20 ft (6.1 m) is the nominal distance used for sensitivity adjustments. The exception is for tests with 0° boom elevation, where the deviation is the horizontal distance from the boom to the closest phase conductor at the point of alarm activation.

Table 10 summarizes PWD alarm performance based on different power line configurations for test sets 1 through 7 when analysis was limited only to tests where the boom extension was at 50% and the boom elevation was such that the tip was approximately at power line level. This configuration approximated the general position used for sensitivity adjustments, so the performance data do not include the adverse effects of having the boom fully retracted and/or at 0° elevation (usually resulting in no alarm), or at full extension and/or 61° elevation (often causing early alarms).

Table 10.—Summary of PWD alarm performance based on power line configuration when analysis is limited only to tests where the boom extension was at 50% and the boom elevation was such that the tip was approximately at power line level

Test set No. and power line configuration		Alarm distance deviation,[1] ft (m) (Boom configuration: boom tip at power line level and boom extension at 50%)	
		SIGALARM Model 210	ASE Model 2100
1	One horizontal line set, 13 kV	6 (1.8)	9 (2.7)
2	One vertical line set, 25 kV	ND	ND
3	Two horizontal line sets on same poles, 13 kV/13 kV	6 (1.8)	10 (3.1)
4	Two horizontal line sets on same poles, 25 kV/13 kV	8 (2.4)	11 (3.4)
5	Two line sets, one vertical over one horizontal on same poles, 25 kV/13 kV	15 (4.6)	12 (3.7)
6	Two horizontal line sets crossing at 90°, 13 kV/13 kV *Left swing*	9 (2.7)	9 (2.7)
6	Two horizontal line sets crossing at 90°, 13 kV/13 kV *Right swing*	4 (1.2)	14 (4.3)
7	Two horizontal line sets crossing at 90°, 25 kV/13 kV *Left swing toward 25-kV lines*	35 (10.7)	34 (10.4)
7	Two horizontal line sets crossing at 90°, 25 kV/13 kV *Right swing toward 13-kV lines*	37 (11.3)	37 (11.3)

ND No data.
[1]Deviation refers to the average of the magnitudes for (true alarm distance − 20 ft (6.1 m)), where 20 ft (6.1 m) is the nominal distance used for sensitivity adjustments.

In analyzing the foregoing data, no distinct pattern emerges with respect to PWDs failing to alarm or alarming late based only on the influence of power line configuration. An examination of the deviations in alarm distance, however, for the cases where the devices did alarm, shows that accuracy was best for simple single-circuit installations (45%–100% average deviation), but degraded somewhat for multiple circuits on the same poles (65%–138% average deviation). This degradation was slightly greater for installations with different voltage levels and/or both vertical and horizontal conductor arrangements. Performance degraded further for crane operation between two intersecting power line installations (90%–215% average deviation). The worst accuracy was associated with intersecting line installations at different voltages.

When analysis is limited to only the data from tests where the boom position approximated sensitivity adjustment conditions (Table 10), PWD accuracy displays a pattern similar to that in the results for all boom positions. In the limited data case, however, PWD performance seemed to be more sensitive to, and adversely affected by, the presence of two different voltage levels near the crane.

Another behavior seen in the data is the tendency of the SIGALARM PWD to be more accurate when swinging left versus right, with the crane positioned between intersecting power line sets. This may be due to placement of the SIGALARM probe on the left side of the boom.

Crane Boom Configurations

Test sets 1 through 7 were also used to examine the effects of different crane boom positions on PWD performance. This evaluation included the 284 individual tests described previously, of which 108 tests with 0° boom should have produced no alarm and the remaining 176 should have ideally caused the PWDs to alarm at approximately 20 ft (6.1 m) from the power lines. Table 11 details the results for these tests when grouped according to boom angle and boom extension. The table lists the number of no alarm, early alarm, and late alarm cases for each boom parameter individually, as well as the average of alarm distance deviations from 20 ft (6.1 m) (magnitudes) for the tests that caused alarm activations.

Table 11.—Summary of PWD alarm performance based on crane boom elevation angle or extension

Crane boom parameter	SIGALARM Model 210		ASE Model 2100	
	Alarm record: (a) No alarm[1] (b) Late alarm[1] (c) Early alarm[2]	Alarm distance deviation,[3] ft (m)	Alarm record: (a) No alarm[1] (b) Late alarm[1] (c) Early alarm[2]	Alarm distance deviation,[3] ft (m)
0° boom elevation	(a) NA (b) NA (c) 11 of 54	40.5 (12.3)	(a) NA (b) NA (c) 4 of 54	28.5 (8.7)
Boom elevation placing boom tip approximately level with power lines	(a) 0 of 34 (b) 2 of 34 (c) 32 of 34	22.4 (6.8)	(a) 0 of 34 (b) 0 of 34 (c) 34 of 34	27.5 (8.4)
61° boom elevation	(a) 25 of 54 (b) 4 of 54 (c) 25 of 54	20.7 (6.3)	(a) 18 of 54 (b) 2 of 54 (c) 34 of 54	22.6 (6.9)
Minimum boom extension *Boom length 28 ft (8.5 m) from base pivot pin to point sheave pin*	(a) 16 of 18 (b) 0 of 18 (c) 3 of 36	18.7 (5.7)	(a) 17 of 18 (b) 0 of 18 (c) 1 of 36	4.0 (1.2)
50% boom extension *Boom length 48.7 ft (14.8 m) from base pivot pin to point sheave pin*	(a) 4 of 34 (b) 6 of 34 (c) 28 of 52	17.5 (5.3)	(a) 1 of 34 (b) 2 of 34 (c) 32 of 52	16.6 (5.1)
100% boom extension *Boom length 69.4 ft (21.2 m) from base pivot pin to point sheave pin*	(a) 5 of 36 (b) 0 of 36 (c) 37 of 54	31.1 (9.5)	(a) 0 of 36 (b) 0 of 36 (c) 39 of 54	31.9 (9.7)

NA Not applicable.

[1] The denominator in this entry refers to tests in the category listed, where an alarm should have occurred at approximately 20 ft (6.1 m) from the power lines.

[2] The denominator in this entry refers to all tests in the category listed.

[3] Deviation refers to the average of the magnitudes for (true alarm distance – 20 ft (6.1 m)), where 20 ft (6.1 m) is the nominal distance used for sensitivity adjustments. The exception is for tests with 0° boom elevation, where the deviation is the horizontal distance from the boom to the closest phase conductor at the point of alarm activation.

Based on the results presented in Table 11, the PWDs were most likely to fail to alarm when the crane boom was operated at minimum extension (89%–94% failed to alarm) and/or at 61° elevation angle (33%–46% failed to alarm). For the tests where the devices did alarm, the accuracy was poorest at 100% boom extension (160% average deviation) and/or 0° boom elevation (145%–205% average deviation).

Other Factors

Use of a PWD as an Early Warning Device

Test sets 8 and 9 emulated the use of a PWD to provide an "early warning" of a power line(s) for a crane traveling into an unfamiliar work area. With the crane in travel configuration and both PWDs set at maximum sensitivity, the SIGALARM Model 210 alarmed at 75–82 ft (22.9–25.0 m) from a single set of 13-kV power lines (depending on approach angle) and 88 ft (26.8 m) from two intersecting sets of 13-kV power lines. These alarm distances should be sufficient for an operator to take preventive measures before the crane is in a position from which it could contact nearby power lines. The ASE Model 2100 allowed the crane, in travel configuration, to approach and travel under the single set of 13-kV lines without alarm activation, but did alarm at a point 20 ft (6.1 m) from the two intersecting sets of 13-kV power lines.

Effect of Connecting the Crane Chassis to a Grounded Power System Neutral

Test set 10 examined the effect of directly connecting the crane chassis to the solidly grounded neutral of a three-phase overhead power line system (all other testing had no deliberate, direct connection of the chassis to the power system neutral). For the test conditions used, the PWDs showed no significant difference in performance due to connection of the crane chassis to a grounded power system neutral.

Effect of Multiple Independent Power Line Circuits

Several tests under test set 7 created conditions under which PWD alarm activation would cease and then resume during one continuous swing of the boom toward a set of power lines. This behavior was documented for two power line sets intersecting at 90°, operating at different voltages, and having opposite phase sequencing. PWD sensitivities were adjusted at 20 ft (6.1 m) from the lower voltage lines, and the boom swing (wherein the alarm dropout occurred) was toward the lower voltage lines while simultaneously moving away from the higher voltage lines. While this behavior was not thoroughly investigated during this research, the several cases in which it was noted clearly illustrate that PWD performance can be compromised when a crane is simultaneously exposed to electric fields from multiple independent overhead power line circuits.

Effect of Power Line Phase Sequence

In test sets 1 through 7, one varied parameter was the phase sequence for the three-phase circuit involved or for one of two circuits when two sets of power lines were used. This is an important issue since there is no convention in distribution power line construction standards for the relative position of specific phases in a three-phase overhead line installation, and any effect of phase sequence on PWD performance will be random. For test sets 1 through 7, tests were run with the sequence as ABC on all circuits, with A-phase nearest the crane (or with A at the top for a vertical conductor arrangement). The tests were then run again with the phase sequence reversed (CBA), either on the one circuit involved or on one of two circuits used. Table 12 details the variations in PWD accuracy due to introduction of CBA phase sequence.

Table 12.—Summary of PWD alarm performance changes due to reversal of power line phase sequence

Test set No. and power line configuration		Change in PWD alarm distance deviation due to introduction of CBA phase sequence[1]	
		SIGALARM Model 210	ASE Model 2100
1	One horizontal line set, 13 kV	16% increase	7% increase
2	One vertical line set, 25 kV	31% increase	7% decrease
3	Two horizontal line sets on same poles, 13 kV/13 kV	47% decrease	60% decrease
4	Two horizontal line sets on same poles, 25 kV/13 kV	25% decrease	3% increase
5	Two line sets, one vertical over one horizontal on same poles, 25 kV/13 kV	41% decrease	63% decrease
6	Two horizontal line sets crossing at 90°, 13 kV/13 kV *Left swing*	36% decrease	36% decrease
6	Two horizontal line sets crossing at 90°, 13 kV/13 kV *Right swing*	36% increase	39% decrease
7	Two horizontal line sets crossing at 90°, 25 kV/13 kV *Left swing toward 25-kV lines*	40% increase	10% increase
7	Two horizontal line sets crossing at 90°, 25 kV/13 kV *Right swing toward 13-kV lines*	29% increase	2% decrease

[1] Change in alarm distance deviation was calculated as ((change in average of alarm distance deviation magnitudes for tests with CBA phase sequence ÷ average of alarm distance deviation magnitudes for tests with ABC phase sequence) × 100). Deviation was defined as follows: (true alarm distance − 20 ft (6.1 m)) for tests producing alarms (except tests with 0° boom elevation), or the horizontal distance from the boom to the closest phase conductor for tests with 0° boom elevation, or 10 ft (3.0 m) for tests that should have caused an alarm but did not. (In the last case, the boom swing was stopped 10 ft (3.0 m) from the closest phase conductor when no alarm occurred.) See "Results" section for details on reversal of three-phase sequence for each test configuration.

Results in Table 12 indicate that the PWDs tested were normally more accurate when two power line circuits on the same poles had opposite phase sequences compared to the same circuits with matching sequences. This was also the pattern for intersecting power line circuits at the same voltage (notwithstanding the SIGALARM's poorer performance for right swings). Opposite phase sequence, however, was more likely to degrade accuracy for intersecting power line circuits operating at different voltages.

Sensitivity Adjustments

Manufacturer's literature for the SIGALARM Model 210 recommends that if the device is to be used as a "proximity warning device" as opposed to an "automatic early power line warning device," the device should be "calibrated" (sensitivity-adjusted) with the boom retracted to the minimum length expected to be used and positioned at a distance from the power lines at least twice that required by 29 CFR or state regulations. The literature adds that if a specific boom extension is to be used for an extended period, the device should be calibrated with the boom at that length [Allied Safety Systems, LLC 2008]. Manufacturer's literature for the ASE Model 2200 (similar to the Model 2100 used in this testing) recommends that the "setpoint" be selected based on electric field strength with the boom at approximately 45° elevation, the boom tip approximately 10 ft (3.0 m) higher than power line level, and a minimum distance from the nearest conductor of no less than 10 ft (3.0 m) [Allied Safety Engineering 2008].

Sensitivity adjustment of PWDs is an issue that directly affects the practical utility of these devices. The test results detailed in this report clearly reveal that many factors can degrade the performance of the PWDs. Foremost among these is change in boom length and elevation angle after a sensitivity adjustment has been made. As a result, selecting the boom position for sensitivity adjustment and, more importantly, recognizing when sensitivity should be readjusted, become important decisions for the user of either PWD tested. For crane operation that involves frequent changes in crane position and/or boom and load configuration, frequent PWD sensitivity adjustments can improve performance. Frequent readjustments, however, could reach a point where it is a distraction and annoyance to the operator and crew. Ultimately, a decision must be reached that balances both PWD accuracy and practicality without compromising personnel safety.

RECOMMENDATIONS FOR FUTURE RESEARCH

Results from this research indicate that several issues need further investigation. PWD performance should be evaluated for crane operation involving only small changes in boom length (less than ± 5 ft (1.5 m) and boom elevation angle (less than ± 8°). This may be a more accurate representation of "typical" crane operation in the field. Two specific areas that warrant more study are crane operation near multiple independent overhead power line circuits and anomalies associated with power line circuit phase sequence reversal. Both of these factors can have an adverse and, from the operator's perspective, unpredictable effect on PWD performance. In addition, research should be conducted to develop and evaluate PWD technologies that can avoid the limitations of electric field measurement that have been highlighted in this report.

REFERENCES

Allied Safety Engineering [2008]. Proximity warning device: ASE model 2200. [http://alliedsafetyeng.com/?mode=pwd]. Date accessed: May 2008.

Allied Safety Systems, LLC [2008]. SIGALARM. [http://www.sigalarminc.com/sigalarm.htm]. Date accessed: May 2008.

ANSI [1995]. American national standard for mobile and locomotive cranes. New York: American National Standards Institute. ANSI/ASME B30.5.

Cawley JC, Homce GT [2003]. Occupational electrical injuries in the United States, 1992–1998, and recommendations for safety research. J Safety Res *34*(3):241–248.

CFR. Code of federal regulations. Washington DC: U.S. Government Printing Office, Office of the Federal Register.

Hipp JE, Henson FD, Martin PE, Phillips GN [1980]. Evaluation of proximity warning devices. Phase I interim report. San Antonio, TX: Southwest Research Institute, pp. 11–13. U.S. Bureau of Mines contract No. J0188082. NTIS No. PB 80-144413.

APPENDIX A.—LIST OF COOPERATORS

Below is a list of the cooperators involved in developing the test plan for the performance evaluation of overhead electrical power line PWDs described in this report.

- Allied Safety Systems, LLC (SIGALARM) (PWD manufacturer): Lance Burney
- Allied Safety Engineering (PWD manufacturer): Irvin Nickerson
- Association of Equipment Manufacturers: Richard Dressler
- International Union of Operating Engineers AFL-CIO: Emmett Russell
- Center for Construction Research and Training (formerly known as the Center to Protect Workers' Rights): Michael McCann, Ph.D., CIH
- Occupational Safety and Health Administration: Kenneth Klouse
- Zachry Construction Corp.: Joseph Collins
- LMK Engineers (NIOSH contractor): Jan Shingler, P.E., Joseph Deane, P.E., and George Gehring, P.E.

www.ingramcontent.com/pod-product-compliance
Lightning Source LLC
Chambersburg PA
CBHW081801170526
45167CB00008B/3277